LINEAR PROGRAMMING

LINEAR PROGRAMMING

S.K. Sharma

CYBER TECH PUBLICATIONS
4264/3, Ansari Road, Daryaganj, New Delhi-110002 (India)
Ph.: 011-23244078. Fax: 011-23280028
E-Mail: cyberpublicationsdelhi@yahoo.com

LINEAR PROGRAMMING

S.K. Sharma

First Edition 2009

Published by :

G.S. Rawat for **Cyber Tech Publications**
4264/3, Ansari Road, Daryaganj, New Delhi-110002 (India)
Ph.: 011-23244078. Fax: 011-23280028
E-Mail: cyberpublicationsdelhi@yahoo.com

Printed at : Asian Offset Printers

PREFACE

This book follows the prescribed syllabus for the subject *"Linear Programming"* of BCA. A major feature of this text is its versatility. Sufficient topics have been included to accommodate students with varies backgrounds. We believe that this occurs because typical students lack the background that most authors of Mathematical Statistics texts assume they have.

The chapters on this book, the students who are weak in these areas. We have found that time devoted to these topics is well spent and pays off when more abstract topics are covered. Another factor that affects readability is the quantity and quality of examples that are relevant to the material being introduced. By providing numerous, clear examples, we hope that we have made this material more accessible to most students.

—*Author*

Contents

1

Trevelling Root Problem of the Variables

INTRODUCTION

Consider the L.P.P Max. $Z = cx$, s.t. $Ax = b$, $x \geq 0$. If x_B is the optimal basic feasible solution and B the optimal basis matrix, we have

$$x_B = B^{-1} b.$$

It is clear that x_B is independent of therefore the change in some component c_j of c will not change x_B, i.e., x_B will always remain basic feasible solution.

Condition of optimality for the solution $\mathbf{x}_B$, is $\Delta_j = c_j - Z_j \leq 0$ for all j not in the basis which is satisfied before any change in c_j. But when c_j changes, the condition of optometry may not be satisfied. The change in the price vector c maybe made in the following two ways:

(i) Variation in $c_j \notin c_B$

(i.e., Change in c_j Which is the Price of the Non-Basic Variable x_j)

Since $\mathbf{x}_B$ is an optimal solution of the L.P.P. (maximization problem)

$\therefore \Delta_j = c_j - Z_j \leq 0$, for all j not in the basis.

If Δc_k is the change in the cost c_k and c_k is not present in c_B (the vector of the costs associated to basic variables) then c_B remains unaltered with this change. Also there is no change in B.

$\therefore Z_j = c_B B^{-1} \alpha_j = c_B Y_j$ also remains unaltered.

If $\mathbf{x}_B$ is still an optimal solution of the given L.P.P. when c_k changes to $c_k + \Delta c_k$ then, $c_j - Z_j$ remains same for all j ($\neq$ k) not in the basis,

$\therefore$ We must have

$$(c_k + \Delta c_k) - Z_k \le 0$$

or $\quad \Delta c_k \le Z_k - c_k \; (= -\Delta_k).$ ...(1)

Hence if $c_k \notin c_B$ changes to $c_k + \Delta c_k$ such that $\Delta c_k \le Z_k - c_k$ $(= -\Delta_k)$, the value of the objective function and the optimal solution of the problem remains unchanged. Note that there is no lower bound to Δc_k.

(ii) Variation in $c_j \in c_B$.

(i.e., Change in c_j Which is the Price of the Basic Variable)

We know that

$$Z_j = c_B B^{-1} a_j = c_B Y_j = \sum_{i=1}^{m} c_{Bi} Y_{ij}.$$

Let Δc_{Bk} be the change in c_{Bk}, the price of the basic variable x_{Bk} then if Z_j^* is the value of Z_j in this solution, then we have

$$Z_j^* = \sum_{\substack{i=1 \\ i \ne k}}^{m} c_{Bi} y_{ij} + (c_{Bk} + \Delta c_{Bk}) y_{kj}$$

$$= \sum_{i=1}^{m} c_{Bi} y_{ij} + y_{kj} \Delta c_{Bk} = Z_j + y_{kj} \Delta c_{Bk}$$

$$\therefore c_j - Z_j^* = c_j - (Z_j + y_{kj} \Delta c_{Bk}) = (c_j - Z_j) - y_{kj} \Delta c_{Bk}.$$

The solution $\mathbf{x}_B$ will remain optimal for the change Δc_{Bk} in c_{Bk} if

$c_j - Z_j^* \le 0$ for all j not in the basis

or $(c_j - Z_j) - y_{kj} \Delta c_{Bk} \le 0$

or $y_{kj}, \Delta c_{Bk} \ge c_j - Z_j$

$$\therefore \Delta c_{Bk} \ge \frac{c_j - Z_j}{y_{kj}}, \text{ for } y_{kj} > 0$$

and $\quad \Delta c_{Bk} \le \dfrac{c_j - Z_j}{y_{kj}}$, for $y_{kj} < 0$. ...(2)

Here the range of Δc_{Bk} (change in the price of c_{Bk} corresponding tooth basis variable x_{Bk}) so that the solution remains optimal is given by

$$\underset{y_{kj} > 0}{\text{Max}} \left[\frac{c_j - Z_j}{y_{kj}} \right] \le \Delta c_{Bk} \le \underset{y_{kj} < 0}{\text{Min}} \left[\frac{c_j - Z_j}{y_{kj}} \right]$$ ...(3)

for all j corresponding to which α_j is not in the optimal basis.

If no $y_{kj} > 0$, there is no lower bound to Δc_{Bk} and if no $y_{kj} < 0$, there is no upper bound to Δc_{Bk}.

To find the change in the value of the objective function.

The value of the objective function for the price vector c is given by

$$\mathbf{Z} = \mathbf{c_B} \cdot \mathbf{x_B} = \sum_{i=1}^{m} c_{Bi} \cdot x_{Bi} .$$

When $c_{Bk} \in c_B$ is changed to $c_{Bk} + \Delta c_{Bk}$, if Z^* is the value of the objective function then

$$Z^* = \sum_{\substack{i=1 \\ i \neq k}}^{m} c_{Bi} \cdot x_{Bi} + (c_{Bk} + \Delta c_{Bk}) x_{Bk}$$

$$= \sum_{i=1}^{m} c_{Bi} \cdot x_{Bi} + x_{Bk} \cdot \Delta c_{Bk} = Z + x_{Bk}\, \Delta\, c_{Bk}.$$

Hence if Δc_{Bk} (change in c_{Bk}) satisfies (3), the solution $\mathbf{x}_B$ will remain optimal and the value of the objective function will be improved by an amount x_{Bk}. Δc_{Bk} where x_{Bk} is the basic variable corresponding to c_{Bk}.

THE PROBLEM OF NEW VARIABLE

If a new variable is introduced in a L.P.P. whose optimal solution has been obtained, then the solution of the problem will remain feasible. Addition of an extra variable x_{n+1} to the problem will introduce an extra column say α_{n+1} to the coefficient matrix A and an extra cost c_{n+1} will be introduced in the price vector c. Thus the addition of this extra variable may affect the optimality of the problem.

For the same basis B the solution $\mathbf{x}_B$ of the original problem will remain optimal (maxima) if

$$c_{n+1} - Z_{n+1} \leq 0$$

$$\Rightarrow c_{n+1} - c_B\, B^{-1}\, \alpha_{n+1} \leq 0. \qquad ...(1)$$

i.e., if (1) is satisfied then the new variable becomes just like a non-basis variable having zero value.

If $c_{n+1} - Z_{n+1} > 0$, then the solution $\mathbf{x}_B$ is no more optimal for the new problem and can be improved by introducing α_{n+1} in the basis.

Here we can start with the last simplex table giving the optimal feasible solution of the original problem by introducing one more column corresponding to the variable x_{n+1}.

Example:

Solve the following L.P.P.

$$Max.\ Z = 3x_1 + 5x_2$$

$$s.t. \quad x_1 + x_3 = 4$$

$$3x_1 + 2x_2 + x_4 = 18$$

$$and \quad x_1, x_2, x_3, x_4 \geq 0.$$

If a new variable x_5 is introduced in the above L.P.P. with price 7, then we have the following problem:

$$Max.\ Z' = 3x_1 + 5x_2 + 7x_5$$

$$s.t. \quad x_1 + x_3 + x_5 = 4$$

$$3x_1 + 2x_2 + x_4 + 2x_5 = 18$$

$$and \quad x_1, x_2, x_3, x_4, x_5 \geq 0.$$

Find the solution of the new L.P.P.

Solution:

Proceeding as usual the successive simplex tables for the original L.P.P. are as follows:

Simplex Table 1

		c_j	3	5	0	0	Mini Ratio
B	c_B	x_B	Y_1	Y_2	Y_3	Y_4	x_B/Y_2
Y_3	0	4	1	0	1	0	–
Y_4	0	18	3	(2)	0	1	18/2 ←
$Z=c_B.x_B=0$		Δ_j	3	5 ↑	0	0 ↓	
Y_3	0	4	1	0	1	0	
Y_2	5	9	3/2	1	0	1/2	
$Z=c_B.x_B=45$		Δ_j	–9/2	0	0	–5/2	

∴ Optimal solution of the given L.P.P. is

$x_1 = 0,\ x_2 = 9,\ x_3 = 4,\ x_4 = 0,$ Max. $Z = 45.$

From the final table the solution of the dual of the given L.P.P. is

$w_1 = 0,\ w_2 = 5/2,$ Min. $Z_D = 45.$

Revised L.P.P. The dual of the new L.P.P. is same as that of the original L.P.P. with one more constraint $w_1 + 2w_2 \geq 7$ which corresponds to the new variable x_5.

The optimal solution of the dual of the original L.P.P. is $w_1 = 0$, $w_2 = 5/2$ which does not satisfy this constraint. Thus we see that the optimal solution of the dual of the given L.P.P. is not optimal for the revised problem and can be improved by introducing $\alpha_5 = [1, 2]$, (column of A corresponding to the new variable introduced) in the basis.

Thus, consider one more column Y_5 in the above table.

$$\text{Since } B^{-1} = \begin{bmatrix} 1 & 0 \\ 0 & 1/2 \end{bmatrix}$$

$$\therefore \quad Y_5 = B^{-1}\alpha_5 = \begin{bmatrix} 1 & 0 \\ 0 & 1/2 \end{bmatrix} \cdot \begin{bmatrix} 1 \\ 2 \end{bmatrix} = \begin{bmatrix} 1 \\ 1 \end{bmatrix}$$

$$\Delta_5 = c_5 - c_B Y_5 = 7 - (0, 5), (1, 1) = 2.$$

We get the following simplex table for revised L.P.P. as

Simplex Table 2

		c_j	3	5	0	0	7	Mini Ratio
B	c_B	x_B	Y_1	Y_2	Y_3	Y_4	Y_5	x_B/Y_5
Y_3	0	4	1	0	1	0	(1)	4/1 ←
Y_2	5	9	3/2	1	0	1/2	1	9/1
$Z'=c_B x_B=45$		Δ_j	−9/2	0	0 ↓	−5/2 ↑	2	
Y_5	7	4	1	0	1	0	1	
Y_2	5	5	1/2	1	−1	1/2	0	
$Z'=c_B x_B=53$		Δ_j	−13/2	0	−2	−5/2	0	

∴ Optimal solution of revised L.P.P. is

$x_1 = 0, x_2 = 5, x_3 = 0, x_4 = 0, x_5 = 4.$

and Max. $Z' = 53$.

THE PROBLEM OF NEW CONSTRAINT

Let a new constraint be introduced to a L.P.P. whose optimal solution has been obtained. Here we assume that the additional constraint does not introduce any new variable with non-zero-price.

Let Z^* be the optimal (maximal) value of the objective function for the new L.P.P. while it was Z for the original problem. Let $Z^* > Z$. Since the new optimal solution satisfies the first m constraints as well as the additional new constraint, therefore, it is also an optimal solution of the original problem which is contradiction to the fact that we already had an optimal solution to the original L.P.P. Here $Z^* \leq Z$. (Max.)

Thus, we have the following two cases:

Case I

If the optimal solution of the original L.P.P. satisfies the new constraint, it is also an optimal solution of the new L.P.P. In this case the adulterant is redundant.

Case II.

If the optimal solution of the original L.P.P. does not satisfy the new constraint, a new optimal solution of the new L.P. problem must be obtained as follows:

To Find the New Optimal Solution of the New Enlarged Problem

Let B and B_1 be the optimal basis of the original and the enlarged L.P. problem respectively. Clearly B_1 is the square matrix of order (m + 1) if B the square matrix of order m.

∴ We can write

$$B_1 = \begin{bmatrix} B & 0 \\ \alpha & \pm 1 \end{bmatrix} \quad ...(1)$$

The last column of B_1 corresponds to the slack, surplus or artificial vector associated with the additional new constraints and α is the row vector of the coefficients, in the new constraint, of the variables which correspond to the vectors in the optimal basis B.

Since B^{-1} exist and is known therefore the inverse of B_1 is given by

$$B_1^{-1} = \begin{bmatrix} B^{-1} & 0 \\ \mp\alpha B^{-1} & \pm 1 \end{bmatrix} \qquad ...(2)$$

Let a_{m+1}, j be the coefficient of x_j in the new (m + 1)th constraint and α_j^* the column vector of the coefficients of x_j in the enlarged problem. Also if Y_j^* and Z_j^* are Y_j and Z_j for the new problem, then we have

$$Y_j^* = B_1^{-1}.\alpha_j^* = \begin{bmatrix} B^{-1} & 0 \\ \mp\alpha B^{-1} & \pm 1 \end{bmatrix} \begin{bmatrix} \alpha_j \\ a_{m+1.j} \end{bmatrix}$$

$$= \begin{bmatrix} Y_j \\ \mp\alpha B^{-1}\alpha_j \pm a_{m+1,j} \end{bmatrix}$$

$$\Rightarrow Y_j^* = \begin{bmatrix} Y_j \\ \mp\alpha Y_j \pm a_{m+1,j} \end{bmatrix}$$

$$\text{Now } Z_j^* = c_{B1}\ Y_j^* = (c_B, c_{B.m+1}) \begin{bmatrix} Y_j \\ \mp\alpha Y_j \pm a_{m+1,j} \end{bmatrix}$$

$$\text{or } Z_j^* = c_{B1}\ Y_j + c_{B,\ m+1}.\ (\pm\ \alpha Y_j \pm a_{m+1,j}) \qquad ...(3)$$

(i) *If slack or surplus variable is introduced in the additional constraint.*

In this case $c_{B,m+1} = 0$

$\therefore$ from (3),

$$Z_j^* = c_B\ Y_j = Z_j$$

$$\therefore c_j - Z_j^* = c_j - Z_j$$

$$\therefore c_j - Z_j^*$$

remains unchanged is this case. Since the optimal solution of the original problem does not satisfy the new constraint, therefore the slack or surplus variable introduced in the new constraint is negative. *Hence we can apply the dual simplex algorithm to find an optimal feasible solution of the new problem.*

(ii) *If the artificial variable is introduced in the additional constraint i.e., if the additional constraint is a perfect equality.* In this case the additional vector is an artificial vector. Now there are two possibilities:

(a) *If the artificial variable in the basic solution is negative,* then assigning a price to the artificial variable we can use the dual simplex algorithm for the removal of the artificial variable from the basis.

(b) *If the artificial variable in the basis solution is positive,* then assigning a price –M to the artificial variable we can use the standard simplex method for the removal of the artificial variable from the basis. It is important to note that in this case $c_j - Z_j$ will be changed.

THE TRAVELLING SALESMAN (ROUTING) PROBLEM

Suppose a salesman wants to visit a certain number of cities.

He knows the distances (or time or cost) of journey between every pair of cities allotted to him. His problem is to choose such a rout which starts from his home city, passes through each city once and only once and returns this home city in the shortest possible distance (or in least time or at least cost).

The above problem may be classified in two forms:

(i) Symmetrical

If the distance (or time or cost) between every pair of cities is independent of the direction of journey the problem is said to be symmetrical.

(ii) Asymmetrical

If for one or more pairs of cities, the distance (or time or cost) changes with the direction, the problem is said to be asymmetrical. For example, it takes longer time while going up-hill from city A to B instead of coming down hill from city B to A. Similarly flying from East to West usually takes longer time than from West to East on account of prevailing winds.

Further we note that for two cities there is only one possible route *i.e.*, there is no choice.

In case of three cities, say A, B and C, one of them (say A) is the home base, there are two possible routes: A → B → C and A → C →

B. For four cities there are 3! = 6 possible routes. In general, there are (n – 1)! possible routes if there are n cities. Thus, practically it is impossible to find the best route by trying each one. That is why the travelling salesman problem is considered as a puzzle by the mathematicians. The best procedure to solve the problem is as if it were an assignment problem. We formulate the problem of travelling salesman in the form of an assignment problem with the additional restriction on his choice of route.

Formulation of a Travelling Salesman Problem as Assignment Problem

Let $x_{ij} = 1$, if the salesman goes directly from city A_j, to c_{ij} to A_j and zero otherwise. Also, let c_{ij} be the distance (or time or cost) from city A_i to city A_j. Then our problem is to minimize $Z = \Sigma\Sigma c_{ij}\, x_{ij}$ with one additional restriction that the x_{ij}'s must be so chosen that no city is visited i_j twice until the tour of all the cities is completed. In particular, he cannot go directly from city A_i to A_i itself. To avoid this possibility in the minimization process we adopt the convention $c_{ij} = \infty$ so that x_{ij} can never be unity when i = j. Also we note that only one $x_{ij} = 1$ for each value of i and j. The distance (or time or cost) matrix for this problem is given in the following table:

		To			
		A_1	A_2	...	A_n
	A_1	∞	c_{12}	...	c_{1n}
	A_2	c_{21}	∞	...	c_{2n}
	.	.	.	.	.
From	.	.	.	.	.
	.	.	.	.	.
	A_n	c_{n1}	c_{n2}	...	∞

We can omit the variable x_{ij} from the problem specification. Our problem is to determine a set of n elements of this matrix, one in each row and one in each column, so as to minimize the sum of the elements determined above.

Note: A problem similar to travelling salesman arises when n items say A_i, i = 1, 2, ... n, are to be produced on a machine in continuation, given that c_{ij} (i, j = 1, 2, ... n) is the setup cost of the machine when item A_i is followed by A_j. Here two additional restrictions are imposed.

One restriction is that we do not follow A_i again by A_i. The other restriction is that we do not produce an item again until all items are produced once.

Solution Procedure

The problem could be solved by assignment technique. In some cases (violation of additional restriction) we use the method of numeration by assignment the next minimum element of matrix in place of zero.

The following example will make the procedure clear.

Example 1:

Solve the travelling salesman problem given by the following data:

$$c_{12} = 20,\ c_{13} = 4,\ c_{14} = 10,\ c_{23} = 5,\ c_{34} = 6$$

$$c_{25} = 10,\ c_{35} = 6,\ c_{45} = 20, \text{ where } c_{ij} = c_{ji}$$

and there is no route between cities i and j if the value for c_{ij} is not given above.

Solution:

Consider the given problem as an assignment problem.

Taking $c_{ij} = \infty$ for i = j, the cost matrix is as follows:

	1	2	3	4	5
1	∞	20	4	10	∞
2	20	∞	5	∞	10
3	4	5	∞	6	6
4	10	∞	6	∞	20
5	∞	10	6	20	∞

If there is no route between cities i and j then we have taken c_{ij} = ∞ to avoid the possibility of going from ith station to jth station. Now we shall solve the problem by usual assignment algorithm. The following tables show the necessary steps for reaching the solution:

∞	15	(0)	4	∞
15	∞	⊗	∞	3
(0)	⊗	∞	⊗	⊗
4	∞	⊗	∞	12
∞	3	⊗	12	∞

∞	12	(0)	1	∞
12	∞	⊗	∞	(0)
(0)	⊗	∞	⊗	⊗
1	∞	⊗	∞	9
∞	(0)	⊗	9	∞

∞	12	(0)	⊗	∞
11	∞	⊗	∞	(0)
⊗	1	∞	(0)	1
(0)	∞	⊗	∞	9
∞	(0)	⊗	8	∞

Hence, the optimum solution of the assignment problem is:

1 → 3, 3 → 4, 4 → 1, 2 → 5, 5 → 2.

But this is not the solution to the travelling salesman problem, as it is not allowed to go from city 4 to 1 without visiting the cities 2 and 5.

We try to find the 'next best' solution which satisfies the additional restriction. The smallest element other than zero is 1. So we try to bring 1 into the solution. Since the element 1 occurs at two places, we shall consider both the cases separately until the acceptable solution is attained.

Make assignment in the cell (3, 2) having the element 1 instead of making zero assignment in the cell (5, 2). After making this assignment we observe that no other assignment can be made in the third row and second column. Consequently make assignment in the cell (5, 4) having element 8, instead of zero assignment. Consequently make assignment in the cell (5, 2).

The new assignment plan is shown in the following table:

To

From		1	2	3	4	5
	1	∞	12	(0)	⊗	∞
	2	11	∞	⊗	∞	(0)
From	3	⊗	(1)	∞	⊗	1
	4	(0)	∞	⊗	∞	9
	5	∞	⊗	∞	(8)	∞

Thus the resulting feasible solution is

1 → 3 → 2 → 5 → 4 → 1 with cost 9.

Again if we make assignment in the cell (3, 5) having the next best element 1 instead of 0 marked in cell (3, 4), then no feasible solution is obtained in terms of zeros or with cost less than 9.

Hence the optimum route is 1 → 3 → 2 → 5 → 4 → 1

The corresponding cost = 4 + 5 + 10 + 20 + 10 = 49.

Example 2:

Given the matrix of setup costs, show how to sequence the production so as to minimize the setup cost per cycle.

		To				
		A_1	A_2	A_3	A_4	A_5
	A_1	∞	2	5	7	1
	A_2	6	∞	3	8	2
From	A_3	8	7	∞	4	7
	A_4	12	4	6	∞	5
	A_5	1	3	2	8	∞

Solution:

Consider the problem as an assignment. Applying the assignment technique, we get the following matrix, showing a solution in terms of marked '□' zero:

		To				
		A_1	A_2	A_3	A_4	A_5
	A_1	∞	1	3	6	(0)
	A_2	4	∞	(0)	6	⊗
From	A_3	4	3	∞	(0)	3
	A_4	8	(0)	1	∞	1
	A_5	(0)	2	⊗	7	∞

The solution to the assignment problem given by above matrix is

$$A_1 \to A_2,\ A_5 \to A_1,\ A_2 \to A_3,\ A_3 \to A_4,\ A_4 \to A_2.$$

This solution indicates to produce the products A_1, then A_5 and then again A_1, without producing the products A_2, A_3 and A_4, which violates the additional restriction of prodding each product once and only once before returning to the first product. So this is not a solution of the travelling salesman problem.

Now we try to find the next best solution which also satisfies the additional restriction. The next minimum element (non-zero) in the matrix is 1. We try to bring 1 in the solution. The cost 1 also occurs at three places.

Start by making unity-assignment in the cell (1, 2) instead of zero assignment in the cell (1, 5). Then no other assignment can be made in the first row and the second column. The best solution of the problem lies in the marked '□' elements as shown in the table.

		A_1	A_2	A_3	A_4	A_5
	A_1	∞	(1)	3	6	⊗
	A_2	4	∞	(0)	6	⊗
From	A_3	4	3	∞	(0)	3
	A_4	8	⊗	1	∞	(1)
	A_5	(0)	2	⊗	7	∞

Thus, the required solution of the problem is

$A_1 \to A_2 \to A_3 \to A_4 \to A_5 \to A_1$.

For this solution, the cost in the reduced matrix is 2.

On the other hand if we select the element 1 in the cell (4, 3) in the solution, then no feasible solution is available in terms of zero or for which the reduced matrix gives the minimum cost less than 2.

Hence, the most suitable sequence is $A_1 \to A_2 \to A_3 \to A_4 \to A_5 \to A_1$.

The minimum setup cost = 2 + 3 + 4 + 5 + 1 = 15.

Example 3:

A salesman estimate that the following would be the cost on his route visiting the six cities as shown.

		To city					
		1	*2*	*3*	*4*	*5*	*6*
	1	∞	*20*	*2*	*27*	*29*	*24*
	2	*21*	∞	*19*	*26*	*31*	*24*
From city	*3*	*26*	*28*	∞	*15*	*36*	*26*
	4	*25*	*16*	*25*	∞	*23*	*18*
	5	*23*	*20*	*23*	*31*	∞	*10*
	6	*27*	*18*	*12*	*35*	*16*	∞

Solution:

Following the usual procedure of assignment algorithms, we obtain Table 12.1 showing an optimum solution indicated by encircled zeros.

This table provides the optimum assignment schedule

1 → 3 → 4 → 2 → 1 and 5 → 6 → 5

with distance according to this optimum route being 101.

Table

	1	2	3	4	5	6
1	∞	0	0	7	2	14
2	0	∞	0	10	8	8
3	6	13	∞	0	14	11
4	4	0	6	∞	0	2
5	8	30	10	21	∞	0
6	13	6	0	26	0	∞

This table does not provide the solution to the travelling salesman problem, as it gives 1 → 3, 3 → 4, 4 → 2, 2 → 1, while city 2 is not allowed to follow city 1 unless city 5 and city 6 processed.

Now we try to find the next best solution which also satisfies this extra restriction. The next minimum (non-zero) element in the matrix is 2. Therefore, we try to bring 2 into the solution. But the element 2 occurs at two places. We shall consider both the cases separately until the acceptable solution is attained,

	1	2	3	4	5	6
1	∞	0	0	7	(2)	14
2	0	∞	0	10	8	8
3	6	13	∞	0	14	11
4	4	0	6	∞	0	2
5	8	30	10	21	∞	0
6	13	6	0	26	0	∞

We start with the element (1, 5) and make the assignment in this cell instead of zero assignment in the cell (1, 3). After making this assignment we see that no other assignment can be made in the first row and third column, and thus the resulting feasible solution will be

$$1 \to 5,\ 5 \to 6,\ 6 \to 3,$$

$$3 \to 4,\ 4 \to 2,\ 2 \to 1.$$

The selected elements for this solution are encircled in Table 12.2. The cost corresponding to this feasible solution is 2.

Again, if we make assignment in the cell (4, 6) instead of (4, 2) then no feasible solution is available in terms of zeros. Hence the best programme is

$$1 \to 5 \to 6 \to 3 \to 4 \to 2 \to 1.$$

The total set-up cost according to this route comes out to be 103.

Solve the travelling salesman problem given by the following data:

$$c_{12} = 20,\ c_{13} = 4,\ c_{14} = 10,$$

$$c_{23} = 5,\ c_{24} = 6,\ c_{25} = 10,$$

$$c_{35} = 6 \text{ and } c_{45} = 20,$$

where $c_{ij} = c_{ij}$

and there is no route between cities i and j if a value for c_{ij} is not shown above.

Given the following matrix of set-up cost, show how to sequence production so as to minimize set-up cost per cycle:

		To				
		A	B	C	D	E
	A	∞	2	5	7	1
	B	6	∞	3	8	2
From	C	8	7	∞	4	7
	D	12	4	6	∞	5
	E	1	3	2	8	∞

A medical representative has to visit five stations A, B, C, D, and E. He does not want to visit any station twice before completing his four of all the stations, and wishes to return to the starting station. Costs of going from one stations to another are given below. Determine the optimal route:

	A	B	C	D	E
A	∞	2	4	7	1
B	5	∞	2	8	2
C	7	6	∞	4	6
D	10	3	5	∞	4
E	1	2	2	8	∞

Example 4:

Find an optimal solution to the following L.P.P.

$Max.\ Z = 15x_1 + 45x_2$

$s.t.\quad x_2 \le 50$

$x_1 + 1.6x_2 \le 240$

$5x_1 + 2.9x_2 \le 162$

and $\quad x_1, x_2\ ^3\ 0.$

If Max. $Z = \sum c_i x_i$ *(i = 1, 2) and* c_2 *is kept fixed at 45, find how much can* c_1 *be changed without affecting the above optimal solution.*

Solution:

Introducing the slack variable x_3, x_4 and x_5, the given problem in standard form for simplex method is as follows:

$$\text{Max. } Z = 15x_1 + 45x_2 + 0.x_3 + 0.x_4 + 0.x_5$$

$$\text{s.t.}\quad 0.x_1 + x_2 + x_3 = 50$$

$$x_1 + 1.6x_2 + x_4 = 240$$

$$5x_1 + 2.0x_2 + x_5 = 162$$

$$\text{and}\quad x_1, x_2, x_3, x_4, x_5 \ge 0.$$

Taking $x_1 = 0 = x_2$, $x_3 = 50$, $x_4 = 240$,

$x_5 = 162$, which is the starting B.F.S.

Solving the given L.P.P. by simple method, we get the following simplex table.

B	c_j	15	45	0	0	0	Min. ratio x_B/Y_2
	c_B	x_B	Y_1	Y_2	Y_3	Y_4	Y_5	
Y_3	0	50	0	(1)	1	0	0	50/1
								←
Y_4	0	240	1	1.6	0	1	0	240/1.6
Y_5	0	162	.5	2	0	0	1	162/2
$Z = c_Bx_B = 0$	Δ_j		15	45	0	0	0	x_B/Y_1
				↑	↓			
Y_2	45	50	0	1	1	0	0	–
Y_4	0	160	1	0	–1.6	1	0	160/1
Y_5	0	62	(.5)	0	–2	0	1	26/.5
								←
$Z = 2250$	Δ_j	15	0	–45	0	1	x_B/Y_3	
		↑				↓		
Y_2	45	50	0	1	1	0	0	50/1
Y_4	0	36	0	0	(2.4)	1	–2	36/2.4
								←
Y_1	15	124	1	0	–4	0	2	–
$Z = 4110$	Δ_j	0	0	15	0	–30		
				↑	↓			
Y_2	45	35	0	1	0	–5/12	5/6	
Y_3	0	15	0	0	1	5/12	–5/6	
Y_1	15	184	1	0	0	5/3	–4/3	
$Z = 4335$	Δ_j	0	0	0	–25/4	–35/2		

From the table we set the optimal solution of the given L.P.P is

$x_1 = 184$, $x_2 = 35$, Max. $Z = 4335$.

To find variation in c_1. Here $c_1 \in c_B$.

From the above simplex table, we have

$c_B = (c_{B1}, c_{B2}, c_{B3}) = (45, 0, 15) = (c_2, c_3, c_1)$

$\therefore c_1 = c_{Bk} = c_{B3} = 15.$

$\therefore$ We have to find the range of variation in $c_1 \in c_B$.

Here we want to find the range Δc_{B3} (change in c_{B3}), which is given by (3), 2.

$$\underset{y_{kj}>0}{\text{Max}}\left[\frac{c_j - Z_j(=\Delta_j)}{y_{kj}}\right] \le \Delta c_{Bk} \le \underset{y_{kj}<0}{\text{Min}}\left[\frac{c_j - Z_j(=\Delta_j)}{y_{kj}}\right]$$

Here k = 3.

Here $y_{kj} = y_{3j} = (1, 0, 0, -4/3)$, out of which y_{31}, y_{32}, y_{33} are not to be taken j = 1, 2, 3 correspond to $\alpha_1, \alpha_2, \alpha_3$ which are in the basis. Thus here we shall consider y_{34} and y_{35} only.

Now $y_{34} = 5/3 > 0$ and $y_{35} = -4/3 < 0$.

$\therefore \Delta c_{B3}$ is given by

$$\text{Max.}\left[\frac{\Delta_4}{y_{34}}\right] \le \Delta c_{B3} \le \text{Min.}\left[\frac{\Delta_5}{y_{35}}\right]$$

or $$\text{Max.}\left[\frac{-25/4}{5/3}\right] \le \Delta c_{B3} \le \text{Min.}\left[\frac{-35/2}{-4/3}\right]$$

or $-15/4 \le \Delta c_{B3} \le 105/8.$

$\therefore$ Range of variation of $c_{B3} = c_1$ is given by

$c_{B3} + (-15/4) \le c_1 \le c_{B3} + 105/8$

or $15 + (-15/4) \le c_1 \le 15 + 105/8$

or $45/4 \le c_1 \le 225/8.$

If we take $c_1 = 45/4$, then proceeding as in above table, we do not get the same optimal solution.

$\therefore$ The variation in c_1 without affecting the above optimal solution is given by $45/4 < c_1 \le 225/8$.

Example 5:

The linear programming problem is

Max $Z = 3x_1 + 5x_2$

s.t. $x_1 + x_2 \le 1$

$2x_2 + 3x_2 \le 1$

and $x_1, x_2 \ge 0.$

Obtain the variations in cj (j = 1, 2) which are permitted without changing the optimal solution.

Solution:

Proceeding as usual, the *final simplex table* giving the optimal solution of the given L.P.P. is as follows.

		c_j	3	5	0	0
B	c_B	x_B	Y_1	Y_2 (β_2)	Y_3 (β_1)	Y_4
Y_3	0	2/3	1/3	0	1	–1/3
Y_2	5	1/3	2/3	1	0	1/3
$Z = c_B x_B = 5/3$		Δ_j	–1/3	0	0	–5/3

From the above table the optimal solution, we have $x_1 = 0$, $x_2 = 1/3$, Max. $Z = 5/3$.

Here $c_B = (c_{B1}, c_{B2}) = (0, 5) = (c_3, c_2)$.

To find variation in c_1. Here c_1 does not belong to c_B.

∴ From (1), 2, the change Δc_1 in c_1, so that the solution remains optimal, is given by

$$\Delta c_1 \le Z_1 - c_1 = -\Delta_1.$$

$$\Rightarrow \Delta c_1 \le 1/3.$$

∴ The range over which c_1 can vary maintaining the optimality of the solution given in above table is

$$-\infty < c_1 \le c_1 + \Delta c_1$$

$$\Rightarrow -\infty < c_1 \le 3 + 1/3$$

$$\Rightarrow -\infty < c_1 \le 10/3.$$

The value of the objective function will not change in this case.

To find variation in c_2. Here $c_2 = c_{Bk} = c_{B2} = 5 \in c_B$

∴ From (3), 2 the range of Δc_{B2} (change in c_{B2}) is given by

$$\underset{y_{kj} > 0}{\text{Max}} \left[\frac{c_j - Z_j}{y_{kj}}\right] \le \Delta c_{Bk} \le \underset{y_{kj} < 0}{\text{Min}} \left[\frac{c_j - Z_j}{y_{kj}}\right].$$

Here k = 2 and $y_{21} = 2/3$, $y_{24} = 1/3 > 0$. Here we cannot consider y_{22} and y_{23} as α_2 and α_3 are in the basis.

Since no $y_{2j} < 0$,

$\therefore$ There is no upper bound to Δc_{B2}.

The change in Δc_{B2} is given by

$$\text{Max.}\left[\frac{\Delta_1}{y_{21}}, \frac{\Delta_4}{y_{24}}\right] \le \Delta c_{b2} < \infty$$

$$\text{Max.}\left[\frac{-1/3}{2/3}, \frac{-5/3}{1/3}\right] \le \Delta c_{B2} < \infty$$

$\Rightarrow -1/2 \le \Delta c_{B2} < \infty$

$\Rightarrow c_{B2} - 1/2 \le c_2 < c_{B2} + \infty$

$\Rightarrow 5 - 1/2 \le c_2 < 5 + \infty$

$\Rightarrow 9/2 \le c_2 < \infty$.

Example 6:

Find an optimal solution to the following L.P. problem

Max. $Z = 3x_1 + 5x_2$

s.t. $x_1 \le 4$

$x_2 \le 6$

$3x_1 + 2x_2 \le 18$ *and* $x_1, x_2 \ge 0$,

what happens to this optimal solution if the objective is changed to

$Z^* = 3x_1 + x_2$.

Solution:

Introducing the slack variables x_3, x_4 and x_5 the given problem in standard form for simplex method is as follows:

Max. $Z = 3x_1 + 5x_2 + 0.x_3 + 0.x_4 + 0.x_5$

s.t $1.x_1 + x_3 = 4$

$0.x_1 + 1.x_2 + x_4 = 6$

$3x_1 + 2x_2 + x_5 = 18$

and $\quad x_1, x_2, \ldots, x_5 \geq 0.$

Taking $x_1 = 0 = x_2$, we get $x_3 = 4$, $x_4 = 6$, $x_5 = 18$, which is the starting B.F.S.

Proceeding as usual we get the following simplex table.

	c_j	3	5	0	0	0	Min. ratio x_B/Y_2	
B	c_B	x_B	Y_1	Y_2	Y_3	Y_4	Y_5	
Y_3	0	4	1	0	1	0	0	–
Y_4	0	6	0	(1)	0	1	0	6/1 ←
Y_5	0	18	3	2	0	0	1	18/2
Z = 0		Δ_j	3	5 ↑	0	0 ↓	0	x_B/Y_1
Y_3	0	4	1	0	1	0	0	4/1
Y_2	5	6	0	1	0	1	0	–
Y_5	0	6	(3)	0	0	–2	1	6/3 ←
Z = 30		Δ_j	3 ↑	0	0	–5	0 ↓	
Y_3	0	2	0	0	1	2/3	–1/3	
Y_2	5	6	0	1	0	1	0	
Y_1	3	2	1	0	0	–2/3	1/3	
Z = 36		Δ_j	0	0	0	–3	–1	

From the above table the

∴ Optimal solution of the given L.P.P. is

$x_1 = 2$, $x_2 = 6$ and Max. Z = 36.

Now the objective function is changed to $Z^* = 3x_1 + x_2$

i.e., c_2 is changed to 1 keeping c_1 fixed.

To find variation in c_2. Here $c_2 \in c_B$, and from last simplex table

$$c_B = (c_{B1}, c_{B2}, c_{B3}) = (0, 5, 3) = (c_3, c_2, c_1)$$

$\therefore c_2 = c_{Bk} = c_{B2} = 5.$

∴ Range of Δc_{B2} is given by

$$\underset{y_{2j}>0}{\text{Max}}\left[\frac{c_j - Z_j(=\Delta_j)}{y_{2j}}\right] \le \Delta c_{B2} \le \underset{y_{2j}<0}{\text{Min}}\left[\frac{c_j - Z_j(=\Delta_j)}{y_{2j}}\right]$$

for all j not in the basis.

$$\Rightarrow \text{Max.}\left[\frac{\Delta_4}{y_{24}}\right] \le \Delta c_{B2} < \infty$$

$$\Rightarrow -3 \le \Delta c_{B2} < \infty$$

$$\therefore 5 - 3 \le c_{B2} < 5 + \infty \text{ or } 2 \le c_2 < \infty$$

$\therefore$ If c_2 is changed to 1, then the optimal solution given in the above table does not remain optimal.

To find the new optimal solution. When Z is changed to Z*.

If the objective function Z is changed to $Z^* = 3x_1 + x_2$ then c_B in the last simplex becomes (0, 1, 3).

$\therefore \Delta_1 = 0\ \Delta_2 = \Delta_3,\ \Delta_4 = c_4 - c_B\ Y_4 = 1,$

$\Delta_5 = c_5 - c_B\ Y_5 = -1.$

Thus from the last of the above table, we get the new table as follows:

		c_j	3	1	0	0	0	Min. ratio x_B/Y_4
B	c_B	x_B	Y_1	Y_2	Y_3	Y_4	Y_5	
Y_3	0	2	0	0	1	(2/3)	–1/3	2/(2/3) ←
Y_2	1	6	0	1	0	1	0	6/1
Y_1	3	2	1	0	0	–2/3	1/3	
$Z^* = 12$	Δ_j	0	0	0	1 ↓	–1 ↑		
Y_4	0	3	0	0	3/2	1	–1/2	
Y_2	1	3	0	1	–3/2	0	1/2	
Y_1	3	4	1	0	1	0	0	
$Z^* = 15$	Δ_j	0	0	–3/2	0	–1/2		

$\therefore$ The solution of the new L.P.P. with changed objective function is $x_1 = 4$, $x_2 = 3$ and Max. $Z^* = 15$.

VARIATION IN THE REQUIREMENT VECTOR b

We know that the condition of optimality for the B.F.S. of a L.P.P. is $\Delta_j = c_j - Z_j \leq 0$.

Since Δ_j does not involve any of b_i if any component b_i of the requirement vector $b = [b_1, b_2, ..., b_m]$ is changed will not affect the conditions of optimality. Hence if any component b_i is changed to $b_i + \Delta b_i$ then the new solution thus obtained will remain optimal. But $x_B = B^{-1} b$ depends on b, therefore any change in b may affect the feasibility of the optimal solution *i.e.*, the optimal solution obtained by changing b may or may not be feasible.

Thus, a change in b_i must be of the magnitude vector b be changed to $b_j + \Delta b_j$, therefore if the new requirement vector is b*, then

$$b^* = [b_1, b_2, \ldots, b_l + \Delta b_l, \ldots, b_m].$$

If x_B^* is the solution of the new L.P.P. obtained by changing b_l to $b_t + Db_l$, then

$x_B^* = B^{-1} b^*$ where B is the optimal basis.

Let $B^{-1} = (\beta_1, \beta_2, \ldots, \beta_m)$

$$= \begin{bmatrix} \beta_{11} & \beta_{12} & \cdots & \beta_{1t} & \cdots & \beta_{lm} \\ \beta_{21} & \beta_{22} & \cdots & \beta_{2t} & \cdots & \beta_{2m} \\ \vdots & \vdots & & \vdots & & \vdots \\ \beta_{il} & \beta_{i2} & \cdots & \beta_{il} & \cdots & \beta_{im} \\ \vdots & \vdots & & \vdots & & \vdots \\ \beta_{ml} & \beta_{m2} & \cdots & \beta_{ml} & \cdots & \beta_{mm} \end{bmatrix}$$

l-th component

Since $b^* = [b_1, b_2, \ldots, b_l + \Delta b_l, \ldots, b_m]$

$= [b_1 + 0, b_2 + 0, \ldots, b_l + \Delta b_l, \ldots, b_m + 0]$

$= [b_1, b_2, \ldots, b_l, \ldots, b_m] + [0, 0, \ldots, \Delta b_l, \ldots, 0]$

$= b + [0, 0, \ldots, \Delta b_l, \ldots, 0]$

$\therefore x_B^* = B^{-1}. b^*$

$= B^{-1}. \{b + [0, 0, \ldots, \Delta b_i, \ldots, 0]\}$

$= B^{-1} b + B^{-1}. [0, 0, \ldots, \Delta b_i, \ldots, 0]$

l-th component

$= x_B + [\beta_1, \beta_2, \ldots, \beta_l, \ldots, \beta_m]. [0, 0, \ldots, \Delta b_l, \ldots, 0]$

$$= x_B + \beta_l . \Delta b_l$$

$$= [x_{B1}, x_{B2}, \ldots, x_{Bl}, \ldots, x_{Bm}] + [\beta_{1l}, \beta_{2l}, \ldots, \beta_{ll}, \ldots, \beta_{ml}] . \Delta b_l .$$

$$= [x_{B1} + \beta_{1l}\, \Delta b_l, \ldots, x_{Bl} + \beta_{ll}\, \Delta b_l, \ldots, x_{Bm} + \beta_{ml} . \Delta b_l] \quad \ldots(1)$$

Now if the solution $x_B{}^*$ is feasible, then

$$x_{Bi} + \beta_{il}\, \Delta b_l \geq 0 \text{ for all } i = 1, 2, \ldots, m$$

$$\Rightarrow \beta_{il}\, \Delta b_l \geq -x_{Bi}$$

$$\therefore \Delta b_l \geq -\frac{x_{Bi}}{\beta_{it}}, \text{ for } \beta_{il} > 0$$

and $\qquad \Delta b_l \leq -\dfrac{x_{Bi}}{\beta_{il}}$. for $\beta_{il} < 0$.

Hence the range of Δ_l so that the optimal solution $x_B{}^*$ also remains feasible is given by

$$\underset{\beta_{il} > 0}{\text{Max}} \left[\frac{x_{Bi}}{\beta_{il}} \right] \leq \Delta b_l \leq \underset{\beta_{il} < 0}{\text{Min.}} \left[-\frac{x_{Bi}}{\beta_{il}} \right]$$

and the new value of the optimal solution is given by (1).

To find the change in the value of the objective function.

The given value of the objection function for a requirement vector is given by

$$Z = c_B\, x_B = \sum_{i=1}^{m} c_{Bi} . x_{Bi} .$$

When b_l is changed to $b_l + \Delta b_l{}^*$ if Z^* is the value of the objective function, then $Z^* = c_B\, x_B$

$$= (c_{B1}, c_{B2}, \ldots, c_{Bm}) . [x_{B1} + b_{1l}\, \Delta_{bl}, \ldots, x_{Bm} + b_{ml}\, \Delta b_l]$$

$$= \sum_{i=1}^{m} c_{Bi} . (x_{Bi} + \beta_{il} . \Delta b\ )$$

$$= \sum_{i=1}^{m} c_{Bi} . x_{Bi} + \sum_{i=1}^{m} x_{Bi} . \beta_{il} . \Delta b_l$$

$$= Z + \sum_{i=1}^{m} x_{Bi} . \beta_{il} . \Delta b_l$$

Hence if Δb_l (change in b_l) satisfies (2), then the solution $x_B{}^*$ given by (1) is also optimal feasible solution. and the value of the objective function is improved by an amount

$$\sum_{i=1}^{m} x_{Bi} \cdot \beta_{il} \cdot \Delta b_l \, .$$

Example:

Given the following L.P.P.

$$Max.\ Z = -x_1 + 2x_2 - x_3$$

$$s.t. \qquad 3x_1 + x_2 - x_3 \le 10$$

$$-x_1 + 4x_2 + x_3 \ge 6$$

$$x_2 + x_3 \le 4$$

$$and \qquad x_1, x_2, x_3 \ge 0.$$

Find the separate range of b_1, b_2 and b_3 (the constants on the right hand sides of the constraints) consistent with the optimal solution.

Solution:

Introducing the slack variables x_4, x_6, surplus variables x_5 and artificial variables x_7, the given L.P.P. reduces to

$$\text{Max. } Z = -x_1 + 2x_2 - x_3 + 0.\,x_4 + 0.x_5 + 0.x_6 - Mx_7$$

$$\text{s.t.} \qquad 3x_1 + x_2 - x_3 + x_4 = 10$$

$$-x_1 + 4x_2 + x_3 - x_5 + x_7 = 6$$

$$x_2 + x_3 + x_6 = 4$$

$$\text{and} \qquad x_1, \ldots, x_7 \ge 0.$$

Proceeding as usual, the successive tables of simplex method are as follows

		c_j	−1	2	−1	0	0	−M	Min. ratio	
B	c_B	x_B	Y_1	Y_2	Y_3	Y_4	Y_5	Y_6	A_1	x_B/Y_2
Y_4	0	10	3	1	−1	1	0	0	0	10/1
A_1	−M	6	−1	(4)	1	0	−1	0	1	6/4 ←
Y_6	0	4	0	1	1	0	0	1	0	4/1
$Z=c_B\ x_B=-6M$		Δ_j	−1 −M	2+4M ↑	−1+M	=	−M	0	0 ↓	x_B/Y
Y_4	0	17/2	13/4	0	−5/4	1	1/4	0		$\frac{17}{2}/\frac{1}{4}$

Y_2	2	3/2	−1/4	1	1/4	0	−1/4	0	$-\frac{5}{2}/\frac{1}{4}$ ←
Y_6	0	5/2	1/4	0	3/4	0	(1/4)	1	
$Z=c_B\ x_B=3$		Δ_j	−1/2	0	−3/2	0	1/2 ↑	0 ↓	
Y_4	0	6	3	0	−2	1	0	−1	
Y_2	2	4	0	1	1	0	0	−1	
Y_5	0	10	1	0	3	0	1	4	
$Z=c_B\ x_B=8$		Δ_j	−1	0	−3	0	0	−2	

We have Optimal solution is $x_1 = 0$, $x_2 = 4$ $x_3 = 0$, Max $Z = 8$.

Here $B = (\alpha_4, \alpha_2, \alpha_5)$, $b = (10, 6, 4)$

$$x_B = (x_4, x_2, x_5) = (x_{B1}, x_{B2}, x_{B3}) = (6, 4, 10),$$

$$\therefore B^{-1} = (\beta_1, \beta_2, \beta_3) = \begin{bmatrix} 1 & 0 & -1 \\ 0 & 1 & 1 \\ 0 & 0 & 4 \end{bmatrix}$$

To find variation in b_1. After changing b_1 to $b_1 + \Delta b_1$, the new requirement vector is given by

$$b_B^* = (10 + \Delta b_1, 6, 4)$$

From relation (2), 3 the range of Δb_1 consistent with the optimal solution is given by

$$\underset{\beta_{il}>0}{\text{Max.}}\left[\frac{-x_{Bi}}{\beta_{il}}\right] \le \Delta b_1 \le \underset{\beta_{il}<0}{\text{Min.}}\left[\frac{-x_{Bi}}{\beta_{il}}\right]$$

Here $b_{11} = 1 > 0$. Since no $b_{il} < 0$,

$\therefore$ there is no upper bound to Δb_1.

$$\therefore \text{We have Max.}\left[-\frac{x_{Bl}}{\beta_{11}}\right] \le \Delta b_1 < \infty.$$

$$\Rightarrow -\frac{6}{1} \le \Delta b_1 < \infty.$$

$\therefore$ Range for b_1 is $-6 + 10 \le b_1 < 10 + \infty$

$\Rightarrow 4 \le b_1 < \infty$

To find variation in b_2.

After changing b_2 to $b_2 + \Delta b_2$, the new requirement vector is $b_B^{**} = (10, 6 + \Delta b_2, 4)$.

From relation (2), 3 the range of Δb_2 consistent with the optimal solution is given by

$$\underset{\beta_{i2} > 0}{\text{Max}}\left[\frac{-x_{Bi}}{\beta_{i2}}\right] \le \Delta b_2 \le \underset{\beta_{i2} < 0}{\text{Min.}}\left[\frac{-x_{Bi}}{\beta_{i2}}\right]$$

Here $\beta_{22} = 1 > 0$. Since no $\beta_{i2} < 0$. $\therefore$ there is no upper bound to Δb_2.

$$\therefore \text{ We have Max.} \left[-\frac{x_{B2}}{\beta_{22}}\right] \le \Delta b_2 < \infty$$

$$\Rightarrow \frac{-4}{1} \le \Delta b_2 < \infty$$

$\therefore$ Range of variation of b_2 is

$$-4 + 6 \le b_2 < \infty + 0$$

$$\Rightarrow 2 \le b_2 < \infty.$$

To find variation in b_3.

After changing b_3 to $b_3 + \Delta b_3$, the new requirement vector is $b_B^{***} = (10, 6, 4 + \Delta_3)$, from relation (2), 3 the range of Δb_3 consistent with the optimal solution is given by

$$\underset{\beta_{i3} > 0}{\text{Max.}}\left[\frac{-x_{Bi}}{\beta_{i3}}\right] \le \Delta b_3 \le \underset{\beta_{i3} < 0}{\text{Min.}}\left[\frac{-x_{Bi}}{\beta_{i3}}\right]$$

Here $\beta_{23} = 1$, $\beta_{33} = 4 > 0$ and $\beta_{13} = -1 < 0$.

$$\therefore \text{ Max.} \left[\frac{-x_{B2}}{\beta_{23}}, -\frac{x_{B3}}{\beta_{33}}\right] \le \Delta b_3 \le \text{Min.}\left[-\frac{x_{B1}}{\beta_{13}}\right]$$

$$\Rightarrow \text{Max.} \left\{-\frac{4}{1}, -\frac{10}{4}\right\} \le \Delta b_3 \le \text{Min.}\left\{\frac{-6}{-1}\right\}$$

$$\Rightarrow -\frac{5}{2} \le \Delta b_3 \le 6$$

$\therefore$ Range of variation of b_3 is

$$4 - \frac{5}{2} \le b_3 \le 4 + 6$$

$$\Rightarrow \frac{3}{2} \le b_3 \le 10.$$

VARIATION IN THE COMPONENT a_{ij} OF THE COEFFICIENT MATRIX A

Let the element a_{lk} of the *l*-th row and k-th column of the coefficient matrix A be changed to $a_{lk} + \Delta a_{lk}$. Now there are two possibilities according as a_{lk} is or not the element of the optimal basis B.

Case I. When a_{lk} is not an Element of Optimal Basis B

If a_{lk} is not an element of the optimal basis B, a change in a_{lk} will not affect B and so B^{-1} remains the same. Thus the change in such a_{lk} will not affect the solution $x_B = B^{-1}b$. Hence the feasibility of the solution x_B is not affected. But the change of a_{lk} may change the optimality of the solution *i.e.*, the solution x_B may not be the optimal solution of the new L.P.P. (obtained by changing a_{lk} to $a_{lk} + \Delta a_{lk}$). Thus we have to find the range of variation of a_{lk} so that the solution still remains optimal.

Since all the column voters except a_k of the matrix A remain unaffected by the variation Δa_{lk} in a_{lk}.

$\therefore \Delta_j = c_j - Z_j \le 0$ for all j ($\neq$ k) not in the basis.

Thus the solution x_B will remain optimal for the new L.P.P. also, if

$$\Delta_k^* = c_k - Z_k^* \le 0 \qquad ...(1)$$

where Δ_k^* and Z_k^* are Δ_k and Δ_k for the new L.P.P.

Let $\quad B^{-1} = (\beta_1, \beta_2, ... ,\beta_l, ... ,\beta_m)$

and $\quad \alpha_k = [a_{lk}, a_{2k}, ..., a_{lk}, ... , a_{mk}]$

If α_k^* is α_k for the new L.P.P. then

$$\alpha_k^* = [a_{lk}, a_{2k}, ... , a_{lk} + \Delta a_{lk}, ..., a_{mk}]$$

$$= [a_{1k} + 0, a_{2k} + 0, ... ,a_{lk} + \Delta_{lk}, ... ,a_{mk} + 0]$$

$$= [a_{1k}, a_{2k}, ... ,a_{lk}, ... ,a_{mk}] + [0, 0, ... ,\Delta a_{lk}, ... ,0]$$

$$= \alpha_k + [0, 0, ... ,\Delta a_{lk}, ... ,0]$$

l-th component

and $\quad Z_k^* = c_B\, B^{-1}\, \alpha_k^*$

$$= c_B.\, B^{-1}\, \{\alpha_k + [0, 0, ... ,\Delta a^{lk}, ... 0]\}$$

$$= c_B.B^{-1}\ \alpha_k + c_B.B^{-1}\ [0, 0, \dots, \Delta a_{lk}, \dots, 0]$$
$$= Z_k + c_B.(\beta_1, \beta_2, \dots, \beta_l, \dots, \beta_m).[0, 0, \dots \Delta a_{lk}, \dots, 0]$$
$$= Z_k + c_B\ (\beta_l\ \Delta a_{lk})$$
$$= Z_k + \Delta a_{lk}.\ c_B\ \beta_l$$
$$= Z_k + \Delta a_{lk}\ (c_{B1}, c_{B2}, \dots, c_{Bl}, \dots, c_{Bm}) \times [\beta_{1l}, \beta_{2l}, \dots, \beta^{ll}, \dots \beta_{ml}]$$
$$= Z_k + \Delta a^{lk} \sum_{i=1}^{m} c_{Bi} \cdot \beta_{il}\ .$$

$\therefore$ From (1), the solution x_B will remain optimal if

$$\Delta^{k*} = c_k - Z_k^* = c_k - Z_k - \Delta a_{lk} \sum_{i=1}^{m} c_{Bi} \cdot \beta_{il} \le 0$$

$$\text{or } \Delta a^{lk} \sum_{i=1}^{m} c_{Bi} \cdot \beta_{il}\ {}^{3}\ c^k - Z^k\ (= \Delta_k)$$

$$\therefore \Delta a_{lk} \ge \frac{\Delta_k}{\sum_{i=1}^{m} c_{Bi}\beta_{il}} \quad \text{for } \sum_{i=1}^{m} c_{Bi}\beta_{il} > 0.$$

$$\text{and} \qquad \Delta a_{lk} \le \frac{\Delta_k}{\sum_{i=1}^{m} c_{Bi}\beta_{il}} \quad \text{for } \sum_{i=1}^{m} c_{Bi}\beta_{il} < 0.$$

Hence the range of Δa_{lk} (change in $a_{lk} \notin B$), so that the solution x_B remains optimal and feasible, is given by

$$\left[\frac{\Delta_k}{\left(\sum_{i=1}^{m} c_{Bi}\beta_{il}\right) > 0}\right] \le \Delta a_{lk} \le \left[\frac{\Delta_k}{\left(\sum_{i=1}^{m} c_{Bi}\beta_{il}\right) < 0}\right] \qquad \dots(2)$$

If $\sum_{i=1}^{m} c_{Bi} \cdot \beta_{il} = 0$, $\Delta\ a_{lk}$ is unrestricted.

If $\sum_{i=1}^{m} c_{Bi} \cdot \beta_{il} > 0$, there is no upper bound to $\Delta\ a_{lk}$,

and if $\sum_{i=1}^{m} c_{Bi} \cdot \beta_{il} < 0$, there is no lower bound to $\Delta\ a_{lk}$.

Note: There is no change in the value of the objective function if D a_{lk} satisfies (2).

Case II. When a_{lk} is an Element of the Optimal Basis B

Since a_{lk} is an element of the optimal basis B then if a_{lk} is changed to $a_{lk} + \Delta a_{lk}$, the optimal basis B will certainly be changed and hence B^{-1} will also be changed.

Consequently $x_B = B-1$ b and $Zj = c_B B^{-1}\alpha_j$ will also change. A change in Z_j may disturb the optimality condition $\Delta_j = c_j - Z_j \le 0$ while a change in x_B may disturb the feasibility of the solution.

Hence our aim is to find the range of variation of a_{lk} so that neither the optimality nor the feasibility of the solution is disturbed.

When the element $a_{lk} \in B$ is changed to $a_{lk} + \Delta a_{lk}$ then let the optimal basis B, the solution x_B and Z_j be represented by B^*, x_B^* and Z^*_j respectively.

Let $\quad B = (b_1, b_2, \ldots, b_m)$ and $\alpha_k = bp$

$\therefore a_{lk} = b_{lp}$ and $a_{lk} + \Delta a_{lk} = b_{lp} + \Delta b_{lp}$

$$\therefore B^* = B = \begin{pmatrix} 0 & 0 & \ldots & 0 & \ldots & 0 \\ 0 & 0 & \ldots & 0 & \ldots & 0 \\ \vdots & \vdots & & \vdots & & \vdots \\ 0 & 0 & \ldots & \Delta b_{lp} & \ldots & 0 \\ \vdots & \vdots & & \vdots & & \vdots \\ 0 & \ldots & 0 & \ldots & 0 & 0 \end{pmatrix} \leftarrow l\text{-th row}.$$

p-th column

$= B + \Delta b_{lp} . 0_{lp}$, where 0_{lp} is the null matrix except for the (l, p)th element which equals to unity

$= B(I + B^{-1} \Delta b_{lp} . 0_{lp})$

$\therefore B^{*-1} = \{B(I + B^{-1} \Delta b_{lp} . 0_{lp})\}-1 = (I + B^{-1} \Delta b_{lp} . 0_{lp})^{-1} . B^{-1}$ ...(3)

$[\because (AB)^{-1} = B^{-1} A^{-1}]$

If $B^{-1} = (\beta_1, \beta_2, \ldots, \beta_m)$,

$$I + B^{-1} \Delta b_{lp} . 0_{lp} = I + B^{-1} \begin{pmatrix} 0 & 0 & \ldots & 0 & \ldots & 0 \\ 0 & 0 & \ldots & 0 & \ldots & 0 \\ \vdots & \vdots & & \vdots & & \vdots \\ 0 & 0 & \ldots & \Delta b_{lp} & \ldots & 0 \\ \vdots & \vdots & & \vdots & & \vdots \\ 0 & \ldots & 0 & \ldots & 0 & 0 \end{pmatrix}$$

$$= I + \xrightarrow[\text{p-th row}]{} \begin{bmatrix} \beta_{11} & \beta_{12} & \cdots & \beta_{1l} & \cdots & \beta_{1m} \\ \beta_{21} & \beta_{22} & \cdots & \beta_{2l} & \cdots & \beta_{2m} \\ \vdots & \vdots & & \vdots & & \vdots \\ \beta_{pl} & \beta_{pl} & \cdots & \beta_{pl} & \cdots & \beta pm \\ \vdots & \vdots & & \vdots & & \vdots \\ \beta_{ml} & \beta_{m2} & \cdots & \beta_{ml} & \cdots & \beta_{mm} \end{bmatrix}$$

l-th column

$$\times \begin{pmatrix} 0 & 0 & \cdots & 0 & \cdots & 0 \\ 0 & 0 & \cdots & 0 & \cdots & 0 \\ \vdots & \vdots & & \vdots & & \vdots \\ 0 & 0 & \cdots & \Delta b_{lp} & \cdots & 0 \\ \vdots & \vdots & & \vdots & & \vdots \\ 0 & \cdots & 0 & \cdots & 0 & 0 \end{pmatrix} \xleftarrow[\text{l-th row}]{}$$

p-th column

$$= \xrightarrow[\text{p-th row}]{} \begin{pmatrix} 1 & 0 & \cdots & 0 & \cdots & 0 \\ 0 & 1 & \cdots & 0 & \cdots & 0 \\ \vdots & \vdots & & \vdots & & \vdots \\ 0 & 0 & \cdots & 1 & \cdots & 0 \\ \vdots & \vdots & & \vdots & & \vdots \\ 0 & 0 & \cdots & 0 & \cdots & 0 \end{pmatrix}$$

p-th column

$$+ \begin{pmatrix} 0 & 0 & \cdots & \beta_{1l}\Delta b_{lp} & \cdots & 0 \\ 0 & 0 & \cdots & \beta_{2l}\Delta b_{lp} & \cdots & 0 \\ \vdots & \vdots & & \vdots & & \vdots \\ 0 & 0 & \cdots & \beta_{pl}\Delta b_{lp} & \cdots & 0 \\ \vdots & \vdots & & \vdots & & \vdots \\ 0 & 0 & \cdots & \beta_{ml}\Delta b_{lp} & \cdots & 0 \end{pmatrix} \xleftarrow[\text{p-th row}]{}$$

p-th column

$$\Rightarrow x_B{}^* = \begin{bmatrix} x_{B1}\, 0 \dfrac{\beta_{1l} \,.\, \Delta b_{lp}}{D} \,.\, x_{Bp} \\ x_{B2} - \dfrac{\beta_{2l} \,.\, \Delta b_{lp}}{D} \,.\, x_{Bp} \\ \vdots \\ \dfrac{1}{D} x_{Bp} \\ x_{Bm} - \dfrac{\beta_{ml} \,.\, \Delta b_{lp}}{D} \,.\, x_{Bp} \end{bmatrix} \qquad \ldots(5)$$

The solution $\mathbf{x}_B{}^*$ is feasible if $\mathbf{x}_B{}^* \geq 0$.

$\therefore$ From (5), the solution $x_B{}^*$ is feasible,

$$\text{if } x_{Bi} - \frac{\beta_{il} \cdot \Delta b_{lp}}{D} \cdot x_{Bp} \leq 0 \qquad ...(6)$$

for all i = 1, 2, ... ,m (i ≠ p)

$$\text{and} \qquad \frac{1}{D} \cdot x_{Bp} \geq 0. \qquad ...(7)$$

Since $x_{Bp} \geq 0$

$\therefore$ (7) will hold of we have

$\therefore$ from (6), (assuming that (8) is true), we have

$x_{Bi}\, D - \beta_{il} .\ \Delta b_{lp}\ x_{Bp} \geq 0$

$\Rightarrow x_{Bi} \cdot (1 + \beta_{pl}\ \Delta b_{lp}) - \beta_{il}\ \Delta b_{pl}\ x_{Bp} \geq 0$

$\Rightarrow x_{Bi} \geq (\beta_{il}\ x_{Bp} - \beta_{pl}\ x_{Bi}) .\ \Delta b_{lp}$ for all $i \neq p$

$$\therefore \Delta\ b_{lp} \leq \frac{x_{Bi}}{\beta_{il} x_{Bp} - \beta_{pl} x_{Bi}}, \text{ for } \beta_{il}\ x_{Bp} - \beta_{pl}\ x_{Bi} < 0$$

$$\text{and} \qquad \Delta\ b_{lp} \geq \frac{x_{Bi}}{\beta_{il} x_{Bp} - \beta_{pl} x_{Bi}}, \text{ for } \beta_{il}\ x_{Bp} - \beta_{pl}\ x_{Bi} < 0.$$

Hence the range of $\Delta\ a_{lk}$ $(= \Delta\ b_{lp}) \in B$, so that the solution remains feasible, is given by

$$\text{Max.} \left[\frac{x_{Bi}}{P_i < 0}\right] \leq \Delta a_{lk} \leq \text{Min}\left[\frac{x_{Bi}}{P_i > 0}\right] \qquad ...(9)$$

where $D = 1 + \beta_{pl}\ \Delta\ \beta_{lp}, = 1 + \beta_{pl}\ \Delta\ a_{lk} > 0$. $Pi = \beta_{il}\ x_{Bp} - \beta_{pl}\ x_{Bi}$.

If no $P_i < 0$, there is no lower bound to $\Delta\ a_{lk}$ and if no $P_i > 0$, there is no upper bound to $\Delta\ a_{lk}$.

To Find the Range of Variation of a_{lk} for the Optimality of the Solution

For optimality of the solution $x_B{}^*$, we must have

$\Delta_j{}^* = c_j - Z_j{}^* \leq 0$

For all j not in the basis.

$$Z_j^* = c_B\, B^{*-1}\, \alpha_j = c_B\, (I + B^{-1}.\ \Delta\, b_{lp}\ 0_{lp})^{-1}.\ B^{-1}\, \alpha_j$$
$$= c_B\, (I + B^{-1}.\ \Delta\, b_{lp}\ 0_{lp})^{-1}.\ Y_j$$

Since $B^{-1}\, \alpha_j = Y_j = [y_{1j}, y_{2j}, \ldots, y_{pj}, \ldots, y_{mj}]$.

Using (4) and simplifying as before as in (1), we have

$$Z_j^* = c_B \begin{bmatrix} y_{1j} - \dfrac{\beta_{1l}\Delta b_{lp}}{D}.y_{pj} \\ y_{2j} - \dfrac{\beta_{2l}\Delta b_{lp}}{D}.y_{pj} \\ \vdots \\ \dfrac{1}{D}y_{pj} \\ \vdots \\ y_{mj} - \dfrac{\beta_{ml}\Delta b_{lp}}{D}.y_{pj} \end{bmatrix}.$$

$$= (c_{B1}, c_{B2}, \ldots, c_{Bp}, \ldots, c_{Bm}). \begin{bmatrix} y_{1j} - \dfrac{\beta_{1l}\Delta b_{lp}}{D}.y_{pj} \\ y_{2j} - \dfrac{\beta_{2l}\Delta b_{pj}}{D} \\ \vdots \\ \dfrac{1}{D}y_{pj} \\ y_{mj} - \dfrac{\beta_{ml}\Delta b_{lp}}{D}.y_{pj} \end{bmatrix} \xleftarrow[\text{p-th row}]{lp.y_{pj}}$$

$$= \sum_{i=1}^{m} c_{Bi}\left(y_{ij} - \frac{\beta_{il}\Delta b_{lp}}{D}y_{pj}\right) + \frac{c_{By}.y_{pj}}{D}$$

$$= \sum_{i=1}^{m} \left(c_{Bi}y_{ij} - \frac{c_{Bi}\beta_{il}\Delta b_{lp}}{D}y_{pj}\right) - c_{Bp}\left(y_{pj} - \frac{\beta_{pl}\Delta b_{lp}}{D}y_{pj}\right) + \frac{c_{Bp}.y_{pj}}{D}$$

$$= \sum_{i=1}^{m} c_{Bi}y_{ij} - \frac{1}{D}\sum_{i=1}^{m} c_{Bi}\beta_{il}\Delta b_{lp}y_{pj} - \frac{1}{D}[c_{Bp}.\ (D - b_{pl}\ \Delta\ b_{lp})\ y_{pj} - c_{Bp}\ y_{pj}]$$

$$= Z_j - \frac{1}{D}\sum_{i=1}^{m} c_{Bi}\beta_{il}\Delta b_{lp} y_{pj}$$

Since $\sum_{i=1}^{m} c_{Bi} y_{ij} = Z_j$ and $D = 1 + \beta_{pl}\,\Delta b_{lp}$

$\therefore$ For optimality of the solution $x_B{}^*$

$\Delta_j{}^* = c_j - Z_j{}^* = c_j - Z_j + \frac{1}{D}\sum_{i=1}^{m} c_{Bi} \cdot \beta_{il}\Delta b_{lp} y_{pj} \le 0 \;\forall$ j not in the basis

$\Rightarrow D\,(c_j - Z_j) + \sum_{i=1}^{m} c_{Bi}\beta_{il}\Delta b_{lp} y_{pj} \le 0$ assuming that $D = 1 + b_{pl}\,\Delta\, b_{lp} > 0$

$$\Rightarrow (1 + b_{pl}\,\Delta b_{lp})\,(c_j - Z_j) + \sum_{i=1}^{m} c_{Bi}\beta_{il}\Delta b_{lp} y_{pj} \le 0$$

$$\Rightarrow \left[\beta_{pl}\,(c_j - Z_j) + \sum_{i=1}^{m} c_{Bi}\beta_{il} y_{pj}\right]\Delta\, b_{lp} \le -\,(c_j - Z_j)$$

$$\Rightarrow Q_j\,\Delta\, b_{lp} \le -\,\Delta_j$$

where $Q_j = b_{pl}\,(c_j - Z_j) + \sum_{i=1}^{m} c_{Bi}\beta_{il} y_{pj}$

$$= b_{pl}\,\Delta_j + \left[\sum_{i=1}^{m} c_{Bi}\beta_{il}\right] y_{pl}$$

$$\therefore \qquad \Delta\, b_{lp} \le \frac{-\Delta_j}{Q_j}, \text{ for } Q_j > 0$$

and $\qquad \Delta\, b_{lp} \ge \frac{-\Delta_j}{Q_j}$, for $Q_j < 0$.

Hence the range of $\Delta\, a_{lk}$ (= $\Delta\, b_{lp}$) so that the solution remains optimal is given by

$$\text{Max.}\left[-\frac{\Delta_j}{Q_j < 0}\right] \le \Delta\, a_{lk} \le \text{Min}\left[\frac{-\Delta_j}{Q_j > 0}\right] \qquad ...(10)$$

for all j not in the basis.

If no $Q_j < 0$, there is no lower bound to $\Delta\, a_{lk}$, and if no $Q_j > 0$, there is no upper bound to $\Delta\, a_{lk}$.

Hence a change Δa_{lk} in a_{lk} (an element of basis matrix B) for the solution to remain feasible and optimal can be made such that (8), (9) and (10) are satisfied.

Example:

Solve the following L.P.P.

$$\text{Max. } Z = 10x_1 + 3x_2 + 6x_3 + 5x_4$$

s.t. $x_1 + 2x_2 + x_4 \le 6$

$3x_1 + 2x_3 \le 5$

$x_2 + 4x_3 + 5x_4 \le 3$

and $x_1, x_2, x_3, x_4 \ge 0.$

Compute the limits for a_{11} and a_{23} so that the new solution remains optimal feasible solution.

Solution:

The given L.P.P. in standard form can be written as follows:

$$\text{Max. } Z = 10x_1 + 3x_2 + 6x_3 + 5x_4 + 0.x_5 + 0.x_6 + 0.\ x_7$$

s.t. $x_1 + 2x_2 + 0.x_3 + x_4 + x_5 = 6$

$3x_1 + 0.x_2 + 2.x_3 + 0.x_4 + x_6 = 5$

$0.x_1 + 1x_2 + 4.x_3 + 5.x_4 + x_7 = 3$

and $x_1, x_2, \ldots, x_7 \ge 0.$

Proceeding as usual the final simplex table is as follows:

		c_j	10	3	6	5	0	0	0
B	c_B	x_B	Y_1	Y_2	Y_3	Y_4	Y_5	Y_6	Y_7
Y_2	3	56/27	0	1	–22/27	0	5/9	–5/27	–1/9
Y_1	10	5/3	1	0	2/3	0	0	1/3	0
Y_4	5	5/27	0	0	26/27	1	–1/9	1/27	2/9
$Z'=c_B\,x_B=643/27$		Δ_j	0	0	–82/27	0	–10/9	–80/27	–7/9

From the above table optimal solution of the given problem is given by

$$x_1 = 5/3,\ x_2 = 56/27,\ x_3 = 0,\ x_4 = 5/27 \text{ and } Z = 643/27.$$

$\therefore B = (Y_2, Y_1, Y_4) = (y_1, y_2, y_3),\ c_B=(c_{B1}, c_{B2}, c_{B3})=(3, 10, 5)$

$x_B = (x_{B1}, x_{B2}, x_{B3}) = (56/27, 5/3, 5/27)\ \alpha_5\ \alpha_6\ \alpha_7$

Since the initial (starting) basis matrix was $(Y_5\ Y_6\ Y_7)$:

from above table

$$B^{-1} = (\beta_1, \beta_2, \beta_3) = \begin{bmatrix} 5/9 & 5/27 & 1/9 \\ 0 & 1/3 & 0 \\ -1/9 & 1/27 & 2/9 \end{bmatrix}$$

i.e., $\beta_{11} = 5/9, \beta_{21} = 0, \beta_{31} = -1/9,$
$\beta_{12} = -5/27, \beta_{22} = 1/3,$
$\beta_{32} = 1/27$ and $\beta_{13} = -1/9, \beta_{23} = 0, \beta_{33} = 2/9.$

To Find Limits of Variation of a_{lk} for the Feasibility of the Solution

From (9) of 4, the range of variation of $a_{lk} = a_{11}$ for the feasibility of the solution is given by

$$\text{Max.}\left[\frac{x_{Bi}}{P_i < 0}\right] \le \Delta\, a_{lk} \le \text{Min.}\left[\frac{x_{Bi}}{P_i > 0}\right] \qquad ...(1)$$

where $D = 1 + \beta_{pl}\,\Delta\,\beta_{lp} \ge 0$, $P_i = \beta_{il}\, x_{Bp} - \beta_{pl}\, x_{Bi}$.
Here $a_{lk} = a_{11} \in B$. Since $a_{11} \in \alpha_1\ (= Y_1) = y_2$
$\therefore a_{lk} = a_{11} = b_{12} = b_{lp}$ *i.e.,* $l = 1$ and $k = 1$, $p = 2$
and $i = 1, 2, 3$. (Since there are only three rows in A)
$x_{B1} = 56/27$, $x_{B2} = 5/3$, $x_{B3} = 5/27$.
$\therefore P_1 = \beta_{11}\, x_{B2} - \beta_{21}\, x_{B1} = (5/9).\,(5/3) - 0 = 25/27 > 0$
$P_2 = \beta_{21}\, x_{B2} - \beta_{21}\, x_{B2} = 0.5/3 - 0 = 0$
$P_3 = \beta_{31}\, x_{B2} - \beta_{21}\, x_{B3} = -1/9.\ 5/3 - 0 = -5/27 < 0.$
$\therefore$ from (1), we have

$$\text{Max.}\left[\frac{x_{B3}}{P_3 < 0}\right] \le \Delta\, a_{11} \le \text{Min.}\left[\frac{x_{B1}}{P_1 > 0}\right]$$

$$\Rightarrow \frac{5/27}{5/27} \le \Delta a_{11} \le \frac{56/27}{25/27}$$

$$\Rightarrow -1 \le \Delta\, a_{11} \le 56/25 \qquad ...(A)$$

Also $D = 1 + \beta_{pl}\,\Delta\, a_{lk} = 1 \ge 0$, holds as $\beta_{pl} = \beta_{21} = 0$.

Again from (10) 4, the range of variation of $a_{lk} = a_{11}$ for the optimality of the solution is given by

$$\text{Max.}\left[\frac{-\Delta_j}{Q_j < 0}\right] \le \Delta a_{lk} \le \text{Min.}\left[\frac{-\Delta_j}{Q_j > 0}\right] \qquad ...(2)$$

where $Q_j = \beta_{pl}\,\Delta_j + \left(\sum_{i=1}^{m} c_{Bi}\beta_{il}\right) y_{pj}$ for all j not in the basis

Here $Q_j = \beta_{21}\,\Delta_j + \left(\sum_{i=1}^{3} c_{Bi}\beta_{il}\right) y_{2j}$

$= \beta_{21}\,\Delta_j + (c_{B1}\,\beta_{11} + c_{B2}\,\beta_{21} + c_{B3}\,\beta_{31})\,y_{2j}$

$= 0 + (3.5/9 + 10.0 - 5.1/9)\,y_{2j} = (10/9)\,y_{2j}$ for all j = 3, 5, 6, 7 not in the basis.

$\therefore Q_3 = 10/9.y_{23} = 10/9.2/3 = 20/27 > 0$

$Q_5 = 10/9.y_{25} = 10/9.0 = 0$

$Q_6 = 10/9.y_{26} = 10/9.1/3 = 10/27 > 0$

$Q_7 = 10/9.y_{27} = 10/9.0 = 0.$

Since no $Q_j < 0$

$\therefore$ There is no lower bound to $a_{lk} - (a_{11})$

$\therefore$ from (2), we have

$$-\infty < \Delta\, a_{11} \le \text{Min.} \left\{-\frac{\Delta_3}{Q_3}, -\frac{\Delta_6}{Q_6}\right\}$$

$$\Rightarrow -\infty < \Delta\, a_{11} \le \text{Min.} \left\{\frac{82/27}{20/27}, \frac{80/27}{10/27}\right\}$$

$\Rightarrow -\infty < \Delta\, a_{11} \le 41/10.$ ...(B)

Since the range for Δa_{11} so that the solution remains optimal and feasible is given (A) and (B) both

$\therefore$ Both (A) and (B) are satisfied

if $-1 \le \Delta\, a_{11} \le 56/25.$

Since $a_{11} = 1$

$\therefore$ limits of variation of a_{11} are

$-1 + 1 \le a_{11} \le 56/25 + 1$

$\Rightarrow 0 \le a_{11} \le 81/25.$

To find limits of variation of a_{23}. Here $a_{23} \notin \alpha_3\,(Y_3)$ which is not in B. From (2), 4 the range of Δa_{lk} (change in $a_{lk} \notin$ B) so that the solution remains optimal and feasible is given by

$$\left[\frac{\Delta_k}{\left(\sum_{i=1}^{m} c_{Bi}\beta_{il}\right) > 0}\right] \le \Delta a_{lk} \le \left[\frac{\Delta_k}{\left(\sum_{i=1}^{m} c_{Bi}\beta_{il}\right)} < 0\right] \quad ...(3)$$

Here $a_{lk} = a_{23} = 2$

$\therefore l = 2,\ k = 3.$

$\Delta_k = \Delta_3 = -82/27$, m = 3 (number of constraints)

$$\sum_{i=1}^{m} c_{Bi}\beta_{i1} = \sum_{i=1}^{3} c_{Bi}\beta_{i2} = c_{B1}\,\beta_{12} + c_{B2}\,\beta_{22} + c_{B3}\,\beta_{32}$$

$= 3. (-5/27) + 10. (1/3) + 5 (1/27) = 80/27 > 0$

$\therefore$ There is no upper bound to $\Delta\, a_{23}$.

$\therefore$ from (3), we have

$$\frac{\Delta_3}{\sum_{i=1}^{3} c_{Bi}\beta_{i2} > 0} \le \Delta\, a_{23} < \infty$$

$$\Rightarrow \frac{-82/27}{80/27} \le \Delta a_{23} < \infty$$

$\Rightarrow -41/40 \le \Delta\, a_{23} < \infty$.

Since $a_{23} = 2$,

$\therefore$ limits of variation of a_{23} are

$-41/40 + 2 \le a_{23} < \infty + 2$

$\Rightarrow 39/40 \le a_{23} < \infty$.

of the given L.P.P. is also the optimal solution of the new L.P.P.

EXERCISES

1. For the L.P.P. Max. $Z = 5x_1 + 3x_2$

   Subject to, $3x_1 + 5x_2 \le 15$

   $5x_1 + 2x_2 \le 10$

   and $x_1, x_2 \ge 0$.

   Find an optimal solution. Hence find how far the component c_1 of the vector c of the function Z = cx can be increased without destroying the optimality of the solution.

2. Find an optimal solution to the following L.P.P.

   Max. $Z = 15x_1 + 45x_2$

   s.t. $x_1 + 16x_2 \le 240$

   $5x_1 + 2x_2 \le 162$

   $x_2 \le 50,\ x_1, x_2 \ge 0$.

   If max. $Z = \sum c_j x_j$, j = 1, 2 and c_2 is fixed at 45, determine how much c_1 can be changed without affecting the above solution.

3. Consider the following table which presents an optimal solution to some L.P.P.

For the above problem assuming that Y_4 and Y_5 were, in that order, in the initial identify matrix basis, calculate the following:

(i) How much can b_1 and b_2 be increase without affecting the optimality and feasibility of the solution?

(ii) How much c_3 can be increased before the present basic solution will no longer be optimal?

B	c_B	x_B	Y_1	Y_2	Y_3	Y_4	y_5
Y_1	2	1	1	0	0.5	4	–0.5
Y_2	3	2	0	1	1	–1	2
		x_j	1	2	0	0	0
	Z = 8	c_j	2	3	1	0	0
		Δ_j	0	0	–3	–5	–5

s.t. $x_1 + 3x_2 - x_3 + 2x_5 - 7$

$-2x_2 + 4x_3 + x_4 = 12$

$-4x_2 + 3x_3 + 8x_5 + x_6 = 10$

and $x_j \geq 0, j = 1, 2, ... ,6.$

For the above L.P.P. calculate the following:

(i) Limits of variations of the costs c_1, c_2, c_3, c_4, c_5 and c_6, for which the optimal solution remains optimal.

(ii) Limits of variation of b_2 without affecting the optimality of the solution.

4. Solve the following L.P.P.

Max. $Z = 3x_1 + 5x_2$

s.t. $x_1 + x_3 = 4$

$3x_1 + 2x_2 + x_4 = 18$

and $x_1, x_2, x_3, x_4 \geq 0.$

Would there be any change in the optimal solution if the additional constraint (1) $x_2 \leq 10$ or (ii) $x_2 \leq 6$ is added to the above L.P.P. In case the optimal solution change find the new optimal feasible solution.

5. (i) Discuss the effect of discrete changes in the requirements (on the right side of the inequalities) for the following L.P.P.

   Max. $Z = 3x_1 + 4x_2 + x_3 + 7x_4$

   subject to

   $8x_1 + 3x_2 + 4x_3 + x_4 \leq 7$

   $2x_1 + 6x_2 + x_3 + 5x_4 \leq 3$

   $x_1 + 4x_2 + 5x_3 + 2x_4 \leq 8,$

   $x_1\ x_2, x_3, x_4 \geq 0.$

   (ii) Discuss the effect of discrete changes in c on the optimality of an optimum basis feasible solution to the above L.P.P.

6. Discuss the effect of adding a new non-negative variable x_8 in the L.P.P. of question 9 (i), on the optimality of its optimum. It is given that the coefficient of x_8 in the constraints of the problem are 2, 7 and 3 respectively, the cost component associated with x_8 being 5.

   (ii) Explain the situation when we have $c_8 = 10$ instead of 5.

7. Consider the L.P.P.

   Mini $Z = x_2 - 3x_3 + 2x_5$

   s.t. $3x_2 - x_3 + 2x_5 \leq 7$

   $-2x_2 + 4x_3 \leq 12$

   $-4x_2 - 3x_3 + 8x_5 \leq 10,$

   $x_2, x_3, x_5 \geq 0.$

   The optimal table is given as follows.

B	c_B	x_B	Y_1	Y_2	Y_3	Y_4	Y_5	Y_6
Y_2	−1	4	2/5	1	0	1/10	4/5	0
Y_3	3	5	1/5	0	1	3/10	2/5	0
Y_6	0	11	1	0	0	−1/2	10	1
		x_j	0	4	5	0	0	11
$Z'=c_B\ x_B=11$		c_j	0	1	−3	0	2	0
$Z=-11$		Δ_j	−1/5	0	0	−4/5	−12/5	0

   (a) Formulate the dual problem for this primal problem.

(b) What are the optimal values of dual variables?

(c) How much c_5 be decreased before Y_5 goes in to basis?

(d) How much can the 7 in first constraint be increased before the basis would change?

8. Solve the following L.P.P. and find how much can c_1 be changed without affecting the optimal conditions.

$$\text{Max. } Z = 15x_1 + 45x_2 = c_1 x_1 + c_2 x_2$$

s.t. $x_1 + 1.6x_2 \leq 240$

$.5x_1 + 2x_2 + x_3 - 162,$

$x_2 + x_4 = 50,$

$x_1, x_2, x_3, x_4 \geq 0.$

9. Consider the following optimal simplex table for a maximization problem (with all constraints of ≤ type), where x_4 is slack and a_1 is an artificial variable. Let a new variable $x_5 \geq 0$ be introduced in the problem with a cost 30 assigned to it in the objective function.

Also suppose that the coefficients of x_5 in the two constraints are 5 and 7 respectively.

B	c_B	x_B	Y_1	Y_2	Y_3	Y_4	A_1
Y_2	12	8/5	0	1	–1/5	2/5	–1/5
Y_1	5	9/5	1	0	7/5	1/5	2/5
Z = 141/5		Δ_j	–3	0	–3/5	–29/5	–M+2/ 5

Discuss the effect of this new variable on the optimality of the given problem.

10. Given that the problem: Max. Z = cx such that Ax = b, x ≥ 0, has an optimal solution; can one obtain a linear programming problem which has an unbounded solution changing b along?

2

Mathematical Description of L.P.P.

INTRODUCTION

A general linear programming problem can be stated as:

Find $x_1, x_2, \ldots x_n$ which optimize the linear function

$$Z = c_1 x_1 + c_2 x_2 + \ldots + c_n x_n \qquad \ldots(1)$$

subject to the constraints

$$\left.\begin{array}{l} a_{11}x_1 + a_{12}x_2 \ldots + a_{1n}x_n (\leq = \geq) b_1 \\ a_{21}x_1 + a_{22}x_2 + \ldots + a_{2n}x_n (\leq = \geq) b_2 \\ \ldots \ldots \ldots \ldots \ldots \ldots \ldots \ldots \ldots \ldots \ldots \\ \ldots \ldots \ldots \ldots \ldots \ldots \ldots \ldots \ldots \ldots \ldots \\ a_{m1}x_1 + a_{m2}x_2 + \ldots + a_{mn}x_n (\leq = \geq) b_m \end{array}\right\} \qquad \ldots(2)$$

and the non-negative restrictions

$$x_1, x_2, \ldots, x_n \geq 0,$$

where a_{ll} $a_{11}, a_{12}, \ldots, a_n, b_1, b_2, \ldots, b_m, c_1, c_2, \ldots, c_n$ are constants and $x_1, x_2, \ldots x_n$ are variables.

The function Z given in (1) is called the *objective function,* the conditions given in (2) are called the linear constraints and the conditions given in (3) are called the non-negative restrictions of the linear programming problem.

This linear programming problem may be stated in matrix form as follows:

Find $x_1, x_2, \ldots, x_n$, so as to optimize

$$Z = c\,x$$

subject to A x (≤ = ≥) b

and $x \geq 0$,

where $A = [a_{ij}]_{m \times n}$ is called the coefficient matrix.

$c = [c_1, c_2, \ldots, c_n]$ is a row matrix known as price vector,

$$b = \begin{bmatrix} b_1 \\ b_2 \\ \ldots \\ \ldots \\ b_m \end{bmatrix}$$ is a column matrix called the requirement vector,

$$x = \begin{bmatrix} x_1 \\ x_2 \\ \ldots \\ \ldots \\ x_n \end{bmatrix}$$ is a column matrix of variables

and 0 is a null matrix of the type $n \times 1$.

Note: In a general linear programming problem it is assumed that the number of rows in the coefficient matrix A is less than the number of its columns.

MATHEMATICAL FORMULATION OF A LINEAR PROGRAMMING PROBLEM

To make use of the linear programming technique it is important to recognize a problem which can be handled by linear programming and then to formulate its mathematical model.

Working Rule to Form a Linear Programming Problem

The following steps are helpful in the formulation of a linear programming problem:

1. Identify the variables in LPP and denote them by x_1, x_2, x_3 etc.
2. Identify the objective function and express it as a linear function of the variables x_1, x_2, x_3 etc.
3. Find the type of the objective function. It may be in the form of maximizing profits or minimizing costs.
4. Identify all the constrains and express them as linear inquisitions or equations.

SLACK AND SURPLUS VARIABLES

1. **Slack Variables.** If a constraint has a sign ≤, then in order to make it an equality we have to add some thing positive to the left hand side.

The positive variables which are added to L.H. sides of the constraints to convert them into equalities are called the slack variables.

For example, consider the linear programming problem

Max. $Z = 2x_1 + 3x_2 + 4x_3$

s.t. $x_1 + x_2 = x_3 \le b_1$

$2x_1 + 4x_2 - x_3 \le b_2$ ...(1)

$x_1, x_2, x_3 \ge 0.$

In order to convert constraints (1) into equalities (equations) we add x_4 and x_5 on L.H.S. of (1), them we have

$x_1 + x_2 + x_3 + x_4 = b_1$

and $2x_1 + 4x_2 - x_3 + x_5 = b_2$.

Hence x_4 and x_5 are called the slack variables.

2. **Surplus Variables.** If a constant has a sign ≥, them in order to make it an equality we have to subtract something positive from its L.H.S. The positive which are subtracted from the L.H. sides of the constants to convert them into equalities are called the surplus variables.

For example, consider the linear programming problem

Max. $Z = 2x_1 + 4x_2 + 4x_3$

s.t. $x_1 + x_2 + 2x_3 \le b_1$

$2x_1 + 4x_2 + 6x_3 \ge b_2$ ...(2)

$x_1 - 2x_2 + 4x_3 \ge b_3$

The constraints (2) are inequalities. In order to convert them into equalities we have to add something in the L.H.S. of first and subtract something from the L.H. sides of the second and the third inequalities of (2), *i.e.,* we have following equalities

$x_1 + x_2 + 2x_3 + x_4 = b_1$

$2x_1 + 4x_2 + 6x_3 - x_5 = b_2$

$x_1 - 2x_2 + 4x_3 - x_6 = b_3$

Here, x_4 is the slack variables while x_5 and x_6 are the surplus variables.

SCOPE OF LINEAR PROGRAMMING

Despite the restriction that relationship must be approximately linear, linear programming, as an OR technique, has been recognized and widely used in industry as an effective tool for business problems, during the last three decades.

The first application of linear programming technique was made in 1947 in the U.S. Air Force. After the Second World War, it became necessary to concentrate on the development of a model that would define the optimal and efficient coordination of a nation's energies in the event of a total war. The Air Force officially named this task as *Project SCOOP* (Scientific Computation of Optimum Programmes) and it was in 1947 that George Dantzing and his associates found out a technique for solving military planning problems while working on this project. This technique consisted of representing the various activities of an organization as a *linear programming modern* and arriving at the optimal programme by minimizing a linear objective function. Dantzing developed the 'Simplex Method' for deriving an optimal feasible solution which solved the project SCOOP. Since then, the military has been building linear programming models for crew training, for scheduling of routine maintenance activities, for personnel assignments and for contract bidding, etc.

The use of linear programming in business and industry has been tremendous since 1951. The first and the most widely used industrial applications have been in all phases of the petroleum industry, viz., exploration, production, refining, distribution and pollution control. Here an oil refinery has a choice of buying crude oil from several different sources with different products like aviation fuel, diesel fuel and gasoline with varying quantities. The constraints may be the restrictions an quantity of the caned oil available at a particular source, the capacity of the refinery to produce a particular product etc.

The second must eater user of linear programming is the food processing industry.

Metal worrying industries use linear programming for loop loading and for determining the cleric between the producing all buying a particular pat.

SOME BASIC CONCEPT AN MATRICAL AND LINEAR ALGEBRA

Consider the system of linear equations

$$3_x + 4_y - 3_z = 5,$$
$$6_x + 8_y - 3_z = 1,$$
$$4_x - 2_y + z = 2.$$

Here x, y and z are unknowns and their coefficient are all numbers arranging the coefficients in the order in which they occur in the equations and enclosing them in square brackets. We obtain a rectangular array of the form

$$\begin{bmatrix} 2 & 9 & 7 \\ 3 & 4 & -3 \\ 6 & 8 & -3 \\ 4 & -2 & 1 \end{bmatrix}.$$

This rectangular array is an example of a **matrix**. The horizontal lines (→) are called rows or *row vectors*, and vertical lines (↓) are called *columns* or *column vectors* of the matrix. There are 4 rows and 3 column in this matrix. Therefore it is a matrix of the type 4 × 3. The numbers 3, 4, – 3, 2 etc., constituting this matrix are called its **elements**. The difference between a matrix and a number should be clearly understood. A matrix is not a number. It has goes got no numerical value. It is a new thing formed with the help of numbers. It is just an ordered collection of number arranged in the form of a rectangular area. Simply 7 is a number. But in our notation of matrices [7] is a matrix of the type 1 × 1 and we cannot have 7 = [7]. We cannot have a relation of equality between a matrix and a number.

We shall use capital letters (in bold type or in italic type) to denote matrices.

$$\text{Thus } \mathbf{A} = \begin{bmatrix} 5 & 0 & 1 \\ 6 & 1 & 7 \end{bmatrix}_{2\times 3} \text{ and } \mathbf{B} = \begin{bmatrix} 0 & 0 & 0 \\ 0 & 0 & 0 \\ 0 & 0 & 0 \end{bmatrix}_{3\times 3}$$

are both matrices. They are of the type 2 × 3 and 3 × 3 respectively.

Sometimes we also use the brackets ( ) or the double bars, || ||, in place of the square brackets [ ] to denote matrices.

$$\text{Thus } \mathbf{A} = \begin{pmatrix} 2 & 2 \\ 2 & 2 \end{pmatrix}, \mathbf{B} = \begin{pmatrix} 3+5i & 9 \\ -4 & 3-5i \end{pmatrix}, \mathbf{C} = \begin{Vmatrix} 7 & 7 \\ 7 & 7 \end{Vmatrix},$$

are all matrices each of the type 2 × 2.

MATRIX

Definition : *Asset of mn numbers (real or complex) arranged in the form of a rectangular array having m rows and n columns is called an m × n matrix [to be read as 'm by n' matrix].*

An m × n matrix is usually written as

$$A = \begin{bmatrix} a_{11} & a_{12} & \dots a_{1n} \\ a_{21} & a_{22} & \dots a_{2n} \\ a_{31} & a_{32} & \dots a_{3n} \\ \dots & \dots & \dots \\ \dots & \dots & \dots \\ a_{m1} & a_{m2} & \dots a_{mn} \end{bmatrix}.$$

In a compact form the above matrix is represented by A – $[a_{ij}]$, i = 1, 2, ... ,m, j = 1, 2, ... n or simply by $[a_{ij}]_{m\times n}$. We write the general element of the matrix and enclose it in brackets of the type [ ] or of the type ( ).

The numbers a_{11}, a_{12} etc. of this rectangular array are called the **elements** of the matrix. The element a_{ij} belongs to the jth row and the jth column and is sometimes called the $(i, j)^{th}$ element of second suffix j, the number of the column in which the element occurs. In a matrix, the number of rows and the columns need not be equal.

SPECIAL TYPES OF MATRICES

(i) Square Matrix

Definition : *An m × n matrix for which for which m = n (i.e., the number of rows is equal to the number of columns) is called a square matrix of order n.* It is also called an n-rowed square matrix. The elements a_{11}, a_{22}, a_{33}, ... ,a_{nn} are called the diagonal elements and the line along which they lie is called the principal diagonal of the matrix. For example the matrix

$$A = \begin{bmatrix} 0 & 1 & 2 & 3 \\ 2 & 3 & 1 & 0 \\ 5 & 0 & 1 & 0 \\ 0 & 0 & 1 & 2 \end{bmatrix}_{4\times 4}$$

is a square matrix of order 4. The elements 0, 3, 1, 2 constitute the principal diagonal of this matrix.

(ii) Unit Matrix or Identity Matrix

Definition : *A square matrix each of whose diagonal elements is 1 and each of whose non-diagonal elements is equal to zero is called a* ***unit matrix*** *or an* ***identity matrix*** *and is denoted by I. In will denote a unit matrix of order n. Thus a sure matrix* $A = [a_{ij}]$ *is a unit matrix if* $a_{ij} = 1$ *when* $i = j$ *and* $a_{ij} = 0$ *when* $i \neq j$.

(iii) Null Matrix of Zero Matrix

Definition : *The* $m \times n$ *matrix whose elements are all 0 is called the* ***null matrix*** *(or zero matrix) of the type* $m \times n$*. It is usually denoted by 0 or more clearly by* $0_{m,\, n}$*. Often a null matrix is simply denoted by the symbol 0 read as 'zero'.*

(iv) Row Matrices and Column Matrices

Definition : *Any* $1 \times n$ *matrix which has only one row and n columns is called a* ***row matrix*** *or row* ***vector****. Similarly any* $m \times 1$ *matrix which has m rows and only column is a* ***column matrix*** *or column vector.*

For example, $\mathbf{X} = [2\ 7\ -8\ 5\ 11]_{1\times5}$ is a row matrix of the type 1×5 while

$$\mathbf{Y} = \begin{bmatrix} 2 \\ -9 \\ 11 \end{bmatrix}_{3\times1}$$ is a column matrix of the type 3×1.

SUBMATRICES OF A MATRIX

Definition : *Any matrix obtained by omitting some rows and columns from a given* $(m \times n)$ *matrix A is called a submatrix of A.*

The matrix **A** itself is a sub-matrix or **A** as it can be obtained from **A** by omitting no rows or columns.

A squares submatrix of a square matrix **A** is called a **principal submatrix,** if its diagonal elements are also the diagonal elements of the matrix **A**. Principal submatrices are obtained only by omitting corresponding rows and columns.

Example:

The matrix $\begin{bmatrix} 1 & 2 & 3 \\ 0 & 2 & 1 \end{bmatrix}$ is a submatrix of the matrix A =

$\begin{bmatrix} 1 & 2 & 3 & 9 \\ 7 & 11 & 6 & 5 \\ 0 & 2 & 1 & 8 \end{bmatrix}$ as it can be obtained from A by omitting the second row and the fourth column.

EQUALITY OF TWO MATRICES

Definition : *Two matrices* $A = [a_{ij}]$ *and* $B = [b_{ij}]$ *are said to be equal if*

(i) *they are of the same size and*

(ii) *the elements in the corresponding places of the two matrices are the same i.e.,* $a_{ij} = b_{ij}$ *for each pair of subscripts i and j.*

ADDITION OF MATRICES

Definition.

Let **A** *and* **B** *be two matrices of the same type m × n. Then their sum (to be denoted by A + B) is defined to be the matrix of the type m × n obtained by adding the corresponding elements of A and B. Thus if* $A = [a_{ij}]_{m\times n}$ and $B = [b_{ij}]_{m\times n}$, then $A + B = [a_{ij} + b_{ij}]_{m\times n}$.

Note that A + B is also a matrix of the type m × n.

For example, if

$$A = \begin{bmatrix} 3 & 2 & -1 \\ 4 & -3 & 1 \end{bmatrix}_{2\times 3} \text{ and } B = \begin{bmatrix} 1 & -2 & 7 \\ 3 & 2 & -1 \end{bmatrix}_{2\times 3},$$

$$\text{then } A + B = \begin{bmatrix} 3+1 & 2-2 & -1+7 \\ 4+3 & -3+2 & 1-1 \end{bmatrix} = \begin{bmatrix} 4 & 0 & 6 \\ 7 & -1 & 0 \end{bmatrix}_{2\times 3}.$$

Note: It should be noted that addition is defined only for matrices which are of the same size. If two matrices A and B are of the same size, they are said to be conformable for *addition.* It the matrices A and B are not of the same size, we cannot find their sum.

PROPERTIES OF MATRIX ADDITION

(i) *Matrix addition is commutative. If A and B be two m × n matrices, then A + B = B + A.*

(ii) *Matrix addition is associative. If A, B, C be three matrices each of the type m × n then (A + B) + C = A + (B + C).*

(iii) *Existence of additive identify. If 0 be the m × n matrix each of whose elements is zero, then* $A + 0 = A = 0 + A$ *for every* $m \times n$ *matrix A.*

(iv) *Existence of the additive inverse.*

Negative of a Matrix

Definition : *Let* $A = [a_{ij}]_{m \times n}$*. Then the negative of the matrix A is defined as the matrix* $[-a_{ij}]_{m \times n}$ *and is denoted by* $-A$.

The matrix $-A$ is the additive inverse of the matrix A. Obviously, $-A + A = O = A + (-A)$.

Here O is the null matrix of the type $m \times n$. It is identity element for matrix addition.

Subtraction of Two Matrices

Definition : *If A and B are two* $m \times n$ *matrices, then we define* $A - B = A + (-B)$.

Thus, the difference $A - B$ is obtained subtracting from each element of A the corresponding element of B.

(v) *Cancellation laws hold good in the case of addition of matrices i.e.,* **if A, B, C** *are three* $m \times n$ *matrices, then*

$$A + B = A + C \Rightarrow B = C$$

and $$B + A = C + A \Rightarrow B = C.$$

(vi) **The equation $A + X = O$ has a unique solution $X = -A$ in the set of all $m \times n$ matrices.**

MULTIPLICATION OF A MATRIX BY A SCALAR

Definition : *Let A be any* $m \times n$ *matrix and k any complex number called scalar. The* $m \times n$ *matrix obtained by multiplying every element of the matrix A by k is called scalar multiple of A by k and is denoted by kA or Ak. Symbolically, if* $A = [a_{ij}]_{m \times n}$, *then* $kA = Ak = [ka_{ij}]_{m \times n}$.

For example, if $k = 2$ and $A = \begin{bmatrix} 3 & 2 & -1 \\ 4 & -3 & 1 \end{bmatrix}_{2 \times 3}$,

then $$2A = \begin{bmatrix} 2 \times 3 & 2 \times 2 & 2 \times -1 \\ 2 \times 4 & 2 \times -3 & 2 \times 1 \end{bmatrix}$$

$$= \begin{bmatrix} 6 & 4 & -2 \\ 8 & -6 & 2 \end{bmatrix}_{2 \times 3}.$$

Properties of Multiplication of a Matrix by a Scalar

Theorem 1:

If A and B are two matrices each of the type $m \times n$, then $k(A + B) = kA + kB$ i.e., scalar multiplication of matrices distributes over the addition of matrices.

Theorem 2:

If p and q are two scalars and A is any $m \times n$ matrix, then $(p + q) A = pA + qA$.

Theorem 3:

If p and q are two scalars and A is any $m \times n$ matrix, then $p(qA) = (pq) A$.

Theorem 4:

If A be any $m \times n$ matrix and k be any scalar, then

$$(-k) A = -(kA) = k(-A).$$

Theorem 5:

If A be any $m \times n$ matrix, then

(i) $1 A = A$,

(ii) $(-1) A = -A$.

Theorem 6:

If A and B are two $m \times n$ matrices, then $-(A + B) = -A - B$.

MULTIPLICATION OF TWO MATRICES

Definition : *Let $A = [a_{ij}]_{m\times n}$ and $B = [b_{jk}]_{m\times n}$ be two matrices such that the number of columns in A is equal to the number of rows in B. Then the $m \times p$ matrix $C = [c_{ik}]_{m\times p}$ such that*

$$c_{ik} = \sum_{j=1}^{n} a_{ij} b_{jk}$$ *[Note that the summation is with respect to the repeated suffix j]*

is called the product of the matrices A and B in that order and we write $C = AB$.

The product AB of two matrices A and B exists if and only if the number of columns in A is equal to the number of rows in B. Two such matrices are said to be conformable for multiplication.

The rule of multiplication is row-by-column multiplication.

For example, if $A = \begin{bmatrix} a_{11} & a_{12} \\ a_{21} & a_{22} \\ a_{31} & a_{32} \end{bmatrix}_{3\times2}$, $B = \begin{bmatrix} b_{11} & b_{12} \\ b_{21} & b_{22} \end{bmatrix}_{2\times2}$

then $$AB = \begin{bmatrix} a_{11}b_{11} + a_{12}b_{21} & a_{11}b_{12} + a_{12}b_{22} \\ a_{21}b_{11} + a_{22}b_{21} & a_{21}b_{12} + a_{22}b_{22} \\ a_{31}b_{11} + a_{31}b_{21} & a_{31}b_{12} + a_{32}b_{22} \end{bmatrix}_{3\times2}.$$

Note: If the product AB exists, then it is not necessary that the product BA will also exist. For example if A is a 4 × 5 matrix and B is a 5 × 3 matrix, then the product AB exists while the product BA does not exist.

ORIGIN AND DEVELOPMENT

L.V. Kentrovich, was the first who was interested in the application of mathematical models to solve such business problems. Decision-Making is the most important aspect of any business and industry. With regard to the problems concerning production of commodities, the main questions before the production manager are, which commodity commodities should be produced. In what quantities and by which press or processes. In this connection. Allured Marshal point out that the businessman always studies his production function and his input price, then substitutes one point for another till has costs become minimum possible.

The model formulated by Canter roved is termed as Linear Programming" which is a decision-making technique uncle given constraints on the assumption that the relation ships amongst the variables representing phenomenon happen to be linear Because of its relative simplicity. Linear Programming was the first area of constrained optimization which attract mathematicians.

The *"Simplex Method"* was the first numerical technique developed to solve linear programming problems. It was developed by George B. Dentition in the late summer of 1947. With the help of econometrics Leonid Heroics and T.C. Koopmans, Dantzin finally developed the algorithm which was mistakenly rejected earlier as being very inefficient.

The importance of simplex method was much more recognized when in 1947 J Don Neumann developed the concept of duality. This concept opened many research areas in linear programming and further enhanced its usefulness and problem-solving ability.

Since 1947, numerous methods have been proposed for solving linear programming problems. During 1950's, contributions have been made by E.M.L. Belle, C.E. Lemke, W. Orchard-Hays, P. Wolfe and others. Together with Dantzing, Philip Wolfe proposed in 1959 a decomposition algorithm, where by large-scale linear programming models can be decomposed into smaller parts that can be optimized independently.

A revolutionary breakthrough in solving complex linear programming problems has recently been achieved by Narender Karmarker, an Indian mathematician. His suppositional *Karmarkar algorithm* has stunned the computer and O.R. communities of the globe. This algorithm is far more effective and efficient than the simplex method. It can handle 10^{10} lakh variables at a time, whereas the simplex method could handle only up to 10^{1000} variables, and that too hemonsuming more computer time. Karmarkar's method enables computers to track, in a fraction of time, a greater combinational of tasks than ever before. The real surprise is that the theoretical breakthrough and the practical applicability, both came together in this discovery.

Since the advent of the "simplex method". Special methods have been developed to solve special types of linear programming problems, such as transportation and assignment models. Kantorovich was the first who identified the *transportation problem* as a special type of linear programming problem in 1939; F.L. Hitchcock developed the standard mathematical formulation of the model in 1947. Since then, numerous methods have been proposed to solve the 'Hitchcock' transportation problem. George Dantzing developed the modified distribution method (MODI) 1951. Charges and Cooper introduced in 1954 the North West corner rule for finding an initial basic solution and presented the simplex-based stepping stone method. H. Kuhn proposed the Hungarian method for the assignment problem in 1955.

REQUIREMENTS OF LINEAR PROGRAMMING

Any organization, big or small, has at its disposal men, machines, money and materials, the supply of which may be limited. Supply of resources bring limited, the management must find the best allocation of its resources in order to maximize the profit or minimize the loss or utilize the production capacity to the maximum extent. However, this involves a number of problems which can be overcome by quantitative methods, particularly the linear programming.

In general, linear programming can be used for optimization problems if the following conditions are satisfied:

(i) There must be a well-defined objective function (profit, cost or quantities produced) which is to be either maximized or minimized and which can be expressed as a linear function of decision variables.

(ii) There must be restrictions on the amount or extent of attainment of the objective and these restrictions must be capable of being expressed as linear qualities or inequalities in terms of variables.

(iii) There must be alternative courses of action. For example, a given product may be processed by two different machines and the problem may be as to how much of the product to allocate to which machine.

(iv) Another necessary requirement is that decision variables should be interrelated and non-negative. The non-negativity condition shows that linear programming deals with real-leaf situations for which negative quantities are generally illogical.

(v) As stated earlier, the resources must be in limited supply. For example, if a firm starts producing getter number of a particular product, it must make smaller number of other products as the total production capacity is limited.

PROPERTIES OF MATRIX MULTIPLICATION

(i) *Matrix multiplication is associative, if convertibility is assured; i.e., $A(BC) = (AB)C$ if A, B, C are $m \times n$, $n \times p$, $p \times q$ matrices respectively.*

(ii) *Multiplication of matrices is distributive with respect to addition of matrices i.e.,*

$$A(B + C) = AB + AC,$$

where A, B, C are any three, $m \times n$, $n \times p$, $n \times p$ matrices respectively.

(iii) *The multiplication of matrices is not always commutative.*

Whenever AB = BA, the matrices A and B are said to commute.

If AB = – BA the matrices A and B are said to anti-commute.

(iv) *If A be any* $m \times n$ *matrix and* $O_{n,p}$ *be an* $n \times p$ *null matrix, then* $AO_{n,p} = 0_{m,p}$ *where* $O_{m,p}$ *is an* $m \times p$ *null matrix.*

Similarly, if $O_{m,n}$ *be an* $m \times n$ *null matrix and A be any* $n \times p$ *matrix,* $O_{m,n} A = O_{m,p}$.

If A be any n-rowed square matrix and O be an -rowed null matrix, then AO = OA = O.

(v) *The equation AB = O does not necessarily imply that at least one of the matrices A and B must be a zero matrix. Or*

The product of two matrices can be a zero matrix while neither of them is a zero matrix.

(vi) *In the case matrix multiplication if AB = O, then it does not necessarily imply that BA = O.*

(vii) *If A be an* $m \times n$ *matrix,* I_n *denotes the n-rowed unit matrix, it can be easily seen that*

$$AI_n = A = I_m A.$$

DETERMINANT OF A SQUARE MATRIX

Let A be any square matrix. The determinant formed by the elements of A is said to be the determinant of matrix A. This is denoted by |A| or det A. Since in a determinant the number of tows is equal to the number of columns, therefore only square matrices can have determinants.

Hence, if $A = \begin{bmatrix} 12 & 0 & 13 \\ 15 & 12 & 11 \\ 13 & 11 & 14 \end{bmatrix}$, then $\det A = |A| = \begin{bmatrix} 12 & 0 & 13 \\ 15 & 12 & 11 \\ 13 & 11 & 14 \end{bmatrix}$.

Difference Between a Matrix and a Determinant

(i) A matrix "A" cannot be reduced to a number whereas the determinant can be reduced to a number.

(ii) The number of rows may or may not be equal to number of columns in a matrix while in a determinant the number of rows is equal to the number of columns.

(iii) Interchanging the rows and column, a different matrix is formed while in a determinant, an interchange of rows and columns does not change the value of the determinant.

NON-SINGULAR AND SINGULAR MATRICES

Definition : *A square matrix A is said to be non-singular or singular according as* $|A| \neq 0$ *or* $|A| = 0$.

TRANSPOSE OF A MATRIX

Definition : *Let* $A = [a_{ij}]_{m\times n}$. *Then the* $n \times m$ *matrix obtained from A by changing its rows into columns and its columns into rows is called the transpose of A and is denoted by the symbol 'A' or* A^T.

The operation of interchanging rows with columns is called transposition. Symbolically, if

$$A = [a_{ij}]_{m\times n},$$

then $$A' = [b_{ji}]_{n\times m},$$

where $$b_{ji} = a_{ij},$$

i.e., the $(j, i)^{th}$ element of A' is the $(i, j)^{th}$ element of A.

For example, the transpose of the 3×4 matrix

$$A = \begin{bmatrix} 1 & 2 & 3 & 4 \\ 2 & 3 & 4 & 1 \\ 3 & 4 & 2 & 1 \end{bmatrix}_{3\times 4}$$ is the 4×3 matrix

$$A' = \begin{bmatrix} 1 & 2 & 3 \\ 2 & 3 & 4 \\ 3 & 4 & 2 \\ 4 & 1 & 1 \end{bmatrix}_{4\times 3}$$

The first row of A is the first column of A'. The second row of A is the second column of A'.

The third row of A is the third column of A'.

Theorems:

If A' and B' be the transposes of A and B respectively, then

(i) $(A')' = A$;

(ii) $(A + B)' = A' + B'$, *A and B being of the same size.*

(iii) $(kA)' = kA'$, *k being any complex number.*

(iv) $(AB)' = B'A'$, *A and B being conformable to multiplication.*

The above law (iv) is called the reversal law for transposes *i.e., the transpose of the product is the product of the transposes taken in the reverse order.*

ADJOINT OR ADJUGATE OF A SQUARE MATRIX

Definition : *Let $A = [a_{ij}]_{n \times n}$ be any $n \times n$ matrix. The transpose B' of the matrix $B = [A_{ij}]_{n \times n}$, where Aij denotes the cofactor of the element a_{ij} in the determinant $|A|$, is called the adjoint of the matrix A and is denoted by the symbol adj A.*

Thus the adjoint of a matrix A is the transpose of the matrix formed by the cofactors of A *i.e.*, if

$$A = \begin{bmatrix} a_{11} & a_{12} & \cdots & a_{1n} \\ a_{21} & a_{22} & \cdots & a_{2n} \\ \cdots & \cdots & \cdots & \cdots \\ a_{n1} & a_{n2} & \cdots & a_{nn} \end{bmatrix},$$

then Adj A = the transpose of the matrix $\begin{bmatrix} A_{11} & A_{12} & \cdots & A_{1n} \\ A_{21} & A_{22} & \cdots & A_{2n} \\ \cdots & \cdots & \cdots & \cdots \\ A_{n1} & A_{n2} & \cdots & A_{nn} \end{bmatrix}$

$$= \text{the matrix} \begin{bmatrix} A_{11} & A_{21} & \cdots & A_{n1} \\ A_{12} & A_{22} & \cdots & A_{n2} \\ \cdots & \cdots & \cdots & \cdots \\ A_{1n} & A_{2n} & \cdots & A_{nn} \end{bmatrix}.$$

Note: Sometimes the adjoint of a matrix is also called the adjudged of that matrix.

Theorem:

If A is any square matrix of order n, then A (adj A) = |A| In = (adj A) A.

INVERTIBLE MATRICES: INVERSE OR RECIPROCAL OF A MATRIX

Definition : *Let A be any n-rowed square matrix. Then a matrix B, if it exists, such that*

$$AB = BA = I_n$$

is called inverse of A.

Note: For the products AB, BA to be both defined and be equal, it is necessary that A and B are both square matrices of the same order. Thus, *non-square matrices cannot possess inverse.*

Existence of the Inverse

Theorem 1:

The necessary and sufficient condition for a square matrix A to possess the inverse is that $|A| \neq 0$.

Note : If A be an invertible matrix, then the inverse of A is $\frac{1}{|A|}$ Adj. A. It is usual to denote the inverse of A by A^{-1}.

Thus, $A^{-1} = \frac{1}{|A|}$ Adj. A, provided $|A| \neq 0$.

Thus the necessary and sufficient condition for a matrix to be invertible is that it is non-singular.

Theorem 1(a):

If A, B be two n-rowed non-singular matrices, then AB is also non-singular and

$$(AB)^{-1} = B^{-1} A^{-1},$$

i.e., the inverse of a product is the product of the inverses taken in he reverse order.

Theorem 2:

If A be an $n \times n$ *non-singular matrix, then* $(A')^{-1} = (A^{-1})'$ *i.e., the operations of transposing and inverting are commutative.*

CONDITION FOR A SYSTEM OF n EQUATIONS IN n UNKNOWNS TO HAVE A UNIQUE SOLUTION

Theorem:

If A be an n-rowed non-singular matrix, X be an $n \times 1$ *matrix, B be an* $n \times 1$ *matrix, the system of equation AX = B has a unique solution.*

Proof:

If A be an n-rowed non-singular matrix, the ranks of the matrices

A and [A B] are both n. Therefore the system of equations AX = B is consistent *i.e.,* possesses a solution.

Pre-multiplying both sides of AX = B by A^{-1}, we have

$$A^{-1} AX = A^{-1} B$$

i.e., $IX = A^{-1} B$

i.e., $X = A^{-1} B$

is a solution of the equation AX = B.

To show that the solution is unique, let us suppose that X_1 and X_2 be two solution of AX = B.

Then $Ax_1 = B, AX_2 = B$

$\Rightarrow$ $AX_1 = AX_2$

$\Rightarrow$ $A^{-1}AX_1 = A^{-1}AX_2$

$\Rightarrow$ $IX_1 = IX_2$

$\Rightarrow$ $X_1 = X_2$.

Hence the solution is unique.

WORKING RULE FOR FINDING THE SOLUTION OF THE EQUATION AX = B

Suppose the coefficient matrix A is of the type m × n, *i.e.,* we have m equations in n unknowns.

Write the augmented matrix [A B] and reduce it to a Echelon form by applying only E-row transformations on it. This Echelon form will enable us to know the ranks of the augmented matrix [A B] and the coefficient matrix A. Then the following different cases arise:

Case I.

Rank A < Rank [A B].

In this case the equations AX = B are inconsistent *i.e.,* they have no solution.

Case II.

Rank A = Rank [A B] = r (say).

In this the equations AX = B are consistent *i.e.,* they possess a solution. If r < m, then in the process of reducing the matrix [A B] to

Echelon form, $(m - r)$ equations will be eliminated. The given system of m equations will then be placed by an equivalent system of r equations. From these r equations we shall be able to express the values of some r unknowns in terms of the remaining $n - r$ unknowns which can be given any arbitrarily chosen values.

If $r = n$, then $n - r = 0$, so that no variable is to be assigned arbitrary values and therefore in this case there will be a unique solution.

If $r < n$, then $n - r$ variables can be assigned arbitrary values. So in this case there will be an infinite number of solution. Only $n - r + 1$ solutions will be linearly independent and the rest of the solutions will be linear combinations of them.

If $m < n$, then $r \leq m < n$. Thus in this case $n - r > 0$. Therefore when the number of equations is less than the number of unknowns, the equations will always have an infinite number of solutions, provided they are consistent.

SOME USEFUL RESULTS OF LINEAR ALGEBRA

1. Basic. A subset S of E^n is said to be the basis of E^n if
   (i) S is linearly independent
   (ii) S generates E^n i.e., $E^n = L(S)$.
2. Generating Set. A set of vectors $\alpha_1, \alpha_2, \ldots, \alpha_k \in E^n$ is said to generate E^n if every vector belonging to E^n can be expressed as a linear combination of $\alpha_1, \alpha_2, \ldots \alpha_k$.
3. Any two bases of E^n have same number of elements.
4. Each set of $n + 1$ or more vectors of E^n is L.D.
5. If a set n vectors of E^n spans E^n then S is a basis of E^n
6. If S is a basis of E^n then every vector $\alpha \in E^n$ can be uniquely expressed as a linear combination of vectors of S.

LINEAR PROGRAMMING AND LINEAR PROGRAMMING PROBLEM

Linear Programming

Linear programming is an important optimization (maximization or minimization) technique used in decision making in business and every-day life for obtaining the maximum or minimum values as required of a linear expression subject to satisfying certain number of given linear restrictions.

Linear Programming Problem (L.P.P.)

The linear programming problem is general calls for optimizing (maximizing/minimizing) a linear function of variables called the objective function subject to a set of linear equations and/or linear inquisitions called the constraints or restrictions. The term linear means that all the relations governing the problem are linear and the term *Programming* refers to the process of determining a particular programme or plan of action.

Objective Function

The function which is to be optimized (maximized or minimized) is called an objective function.

Constraints

The system of linear inquisition (or equations) under which the objective function is to be optimized is called constraints.

BASIC REQUIREMENTS OF A LINEAR PROGRAMMING PROBLEM (L.P.P.)

(a) **Well-defined Objective Function.** The objective function of the L.P.P. should be clear and well-defined. The objective may be either to maximize the profit by utilizing the available resources or it may be to minimize the cost by using limited amount of various resources required for production.

(b) **Quantitative Measurement of Elements.** It is necessary that each element may be capability being measured quantitatively. Numerical data must be given so that the relationships among the elements may be considered.

(c) **Presence of Constraints and Restrictions.** The available resources which are to be allocated among various competing activities should be limited.

(d) **Alternate Lines of Action.** There must be different lines of actions available to perform the job.

(e) **Non-negative Restrictions.** All the variables considered for making decisions assume non-negative values.

(f) **Linearity.** A primary requirement of a L.P.P. is that both the objective function and all the constraints must be expressed in terms of linear equations and inequalities.

Linearity implies that the products of variables such as xy, x^2y etc., powers of variables such as x^2, x^3 etc., and combinations of variables of the type ax + b log y etc. are not allowed.

(g) **Additivity.** Additivity means that if it takes x minutes on machine A to make product P and Y minutes to make product Q, then the total time on machine A devotes to this production is x + y, provided the time required to change the machine from product P to product Q is negligible.

(h) **Multiplicativity.** Multiplicativity means that if the cost of production of 1 unit of an item is Rs. 10, then cost of production of x units of the same item is Rs. 10x.

In terms of profit we can say that the profit from selling a given number of units is the profit per unit multiplied by the number of units sold.

(I) **Divisibility.** Divisibility means that the fractional levels of variables must be permissible besides integral values.

(J) **Finite Number of Constraints.** The number of activities involved in a problem should be finite, thus leading to a finite number of constraints in the problem.

Example 1:

A goldsmith manufactures necklaces and bracelets. The total number of necklaces and bracelets that he can handle per day is at most 24. It takes one hour to make a bracelet and half an hour to make a necklace. It is assumed that he can work for a maximum of 16 hours a day. Further the profit an a bracelet is Rs. 300 and the profit on a necklace is Rs. 100. Formulate this problem as a linear programming problem so as to maximize the profit.

Solution:

Suppose the goldsmith manufactures x_1 necklaces and x_2 bracelets per day.

Since the profit an a necklace is Rs. 100 and profit on a bracelet i s

Rs. 300, therefore the total profit Z in Rs. is given by

$$Z = 100x_1 + 300x_2 \qquad ...(1)$$

Since it takes half an hour to make one necklace, so the time required to make x_1 necklaces = (1/2) x; hours.

Again it takes one hour to make one bracelet, so the time required to make x_2 bracelets = 1. x_2 hours = x_2 hours.

Therefore, total time required to make x_1 necklaces and x_2 bracelets

$$= \left(\frac{x_1}{2} + x2\right) \text{ hours.} \qquad ...(2)$$

Since total time available per day is 16 hours, therefore

$$\frac{x_1}{2} + x_2 \le 16 \text{ or } x_1 + 2x_2 \le 32. \qquad ...(3)$$

The total number of necklaces and bracelets that the goldsmith can manufacture in a day is almost 24, so we have

Example 2:

A firm manufactures A, B veld. The profit per unit soled of each product of Rs. 3, Rs. 2 and Rs. 4 respectively. The time required to manufacture one unit of each of the three products and the daily capacity of the two machines P and Q is given 24 the total below.

Machine	*Time per unit (minutes)*			*Machine capacity*
	A	*B*	*C*	*(Minutes day)*
P	*4*	*3*	*5*	*2000*
2	*2*	*2*	*4*	*2500.*

It is required to determine the daily number of unit to be manufactured for each product so as to maximize two profity. How are the firm must manufacture at least to A's, 200 B's and 50 C's but move then 150 A's. It is assumed that all the amounts produced are connected in the martlet.

VECTORS

Definition : *Any ordered n-tuple of numbers is called an n-vector.* By an ordered n-tuple we mean a set consisting of numbers in which the place of each number is fixed. If x_1, x_2, ... x_n be any n numbers, then the ordered n-tuple **X** = (x_1, x_2, ... ,x_n) is called an n-vector. The ordered triad (x_1, x_2, x_3) is called a 3-vector. Similarly (1, 0, 1, – 1) and (1, 8, – 5, 7) are 4-vectors. The n numbers x_1, x_2, ... ,mn are called components of the n-vector **X** = $(x_1, x_2, \ldots, x_n)$. **A** vector may be written either as a row vector or as a *column vector*. If **A** be a matrix of the type m × n, then each row of **A** will be an n-vector and each column of **A** will be an m-vector. **A** vector whose components are all zero is called a zero vector and will be denoted by **O**.

If k be any number and **X** be any vector, then relative to the vector **X**. k is called a scalar.

Algebra of Vectors

Since an n-vector is nothing but a row matrix or a column matrix. Therefore we can develop an algebra of vectors in the same manner as the algebra of matrices.

Equality of Two Vectors

Two n-vectors **X** and **Y** where **X** = $(x_1, x_2, \ldots x_n)$ and **Y** = $(y_1, y_2, \ldots y_n)$ are said to be *equal* if and only if their corresponding components are equal *i.e.,* if $x_i = y_i$, for all i = 1, 2, ... ,n.

For example if

$$X = (1, 4, 7) \text{ and } Y = (1, 4, 7),$$

then $X = Y.$

Multiplication of a Vector by a Scalar (Number)

If k be any number and

$$\mathbf{X} = (x_1, x_2, \ldots x_n),$$

then by definition

$$k\,\mathbf{X} = (kx_1, kx_2, \ldots, kx_n).$$

The vector k **X** is called the scalar multiple of the vector **X** by the scalar k.

Properties of addition and scalar multiplication of vectors.

If **X, Y, Z** be any three n-vectors and p, q be any two numbers, then obviously

(i) $X + Y = Y + X$.

(ii) $X + (Y + Z) = (X + Y) + Z$.

(iii) $p(X + Y) = pX + pY$.

(iv) (p + q) X = pX – qX.

(v) p (qX) = (pq) X.

LINEAR DEPENDENCE AND LINEAR INDEPENDENCE OF VECTORS

Linearly Dependent Det of Vectors

Definition : *A set r n-vectors $X_1, X_2, \ldots X_r$ is said to be linearly dependent if there exist r scalars (numbers) $k_1, k_2, \ldots k_r$ not all zero, such that*

$$k_1 X_1 + k_2 X_2 + \ldots + k_r X_r = O,$$

where, O, denotes the n-vector whose components are all zero.

Linearly Independent Set of Vectors

Definition : *A set r n-vectors $X_1, X_2, \ldots X_r$ is said to be linearly independent if every relation of the type*

$$k_1 X_1 + k_2 X_2 + \ldots + d_r X_r = O$$

implies $k_1 = k_2 = k_3 = \ldots = k_r = 0.$

A VECTOR AS A LINEAR COMBINATION OF VECTORS

Definition : *A vector X which can be expressed in the form*

$$X = k_1 X_1 + k_2 X_2 + \ldots + k_r X_r$$

is said to be a linear combination of the set of vectors $X_1, X_2, \ldots, X_r$.

Here $k_1, k_2, \ldots, k_r$ are any numbers.

The following two results are quite obvious:

(i) *If a set of vector is linearly dependent, then at least one member of the set can be expressed as a linear combination of the remaining members.*

(ii) *If a set of vectors is linearly independent then no member of the set can be expressed as a linear combination of the remaining members.*

BASIC SOLUTION

Consider a system A x = b of m equations in n unknowns (n > m) and r (A) = r (Ab) = m *i.e.,* none of the equations is redundant.

A solution obtained by setting any (n – m) variables to zero is called a basic solution, provided the determinant of the coefficient of the remaining m variables is not zero.

Such m variables (any of them may be zero) are called basic variables and remaining (n – m) zero valued variables are called non-basic variables.

Thus for a solution to be basic, at least n – m variables must be zero. Here we note that the matrix formed by the coefficients of the m basic variables, or say formed by the vectors associated to the basic variables is non-singular as its determinant does not vanish. Hence the vectors associated to the basic variables are L.I.

Thus *a solution in which the vectors associated to m variables are L.I. and remaining n – m variables are zero is called a basic solution.*

Hence a basic solution can be constructed by selecting the m L. I. vectors out of n and setting the variables associated to the remaining (n – m) columns to zero. If B is the matrix of m L.I. vectors of A and x_B is the vector of the corresponding variables (basic variables), then the basic solution is given by

$$Bx_B = b$$

or

$$x_B = B^{-1}b.$$

The number of basic solutions thus obtained will be at the most ^nC_m $= \dfrac{n!}{m!(n-m)!}$ since m vectors out of n can be selected in ^nC_m ways. Note that a B. S. corresponds to some basis. Basic solution are of two types: Non-degenerate basic solution and degenerate basic solutions.

If any of the basic variables vanishes, the solution is called degenerate basic solution. On the other hand, *if none of the basic variables vanishes, the solution is called non-degenerate basic solution.*

Thus a non-degenerate basic solution constant exactly m non-zero and (n – m) zero variables.

Theorem:

A necessary and sufficient condition for the existence and non-degeneracy of all the basic solutions of Ax = b is that every set of m columns of the augmented matrix Ab = [A, b] is L.I.

Proof:

The condition is necessary. Suppose all the basic solution of the system Ax = b exist and are non-degenerate. Therefore, every set of m-column vectors of A is L.I.

Let $\alpha_1, \alpha_2, \dots, \alpha_m$ be one set of m-column vectors of A, then the given system gives

$$\alpha_1 x_1 + \alpha_2 x_2 + \dots + \alpha_m x_m = b.$$

Now since each solution is non-degenerate, $x_1 \neq 0$ and hence by replacement theorem vector b can replace α_1 in the basis $\alpha_1, \alpha_2, \dots \alpha_m$. Thus vectors b, $\alpha_2, \alpha_3, \dots, \alpha_m$ also from a basis and hence they are L.I. Similarly b can replace α_2 as x_2 is also non-zero. Hence α_1, b, $\alpha_3, \dots, \alpha_m$ are also L.I. In a similar way we can show that b with any (m – 1) vectors of A forms a L.I. set. Hence every set of m columns of the augmented matrix Ab = [A, b] is L.I.

The condition is sufficient. Suppose every set of m column vectors of Ab = [A, b] is L.I. Then obviously every set of m-column vectors of A is also L.I. and hence all the basic solutions exist.

Now consider any m-column vectors $\alpha_1, \alpha_2, \dots, \alpha_m$ of A which are L.I. and let $x_1, x_2, \dots, x_m$ be the corresponding basic solution. Then

$$x_1 \alpha_1 + x_2 \alpha_2 + \dots + x_m \alpha_m = b$$

or $$-1.b + x_1 \alpha_1 + x_2 \alpha_2 + \dots + x_m \alpha_m = 0.$$

Now if $x_1 = 0$, then we have

$$-1.b + x_2 a_2 + \dots + x_m \alpha_m = 0,$$

which shows that the vectors b, $\alpha_2, \dots, \alpha_m$ are L.D.

But this is a contradiction as we have already assumed that m vectors of [A, b] are L.I.

Hence $x_1 \neq 0$.

Similarly, we can show that $x_2, \dots x_m$ are also different from zero *i.e.,* the solution is non-degenerate.

In this way, it can be shown that all the basic solution are non-degenerate.

Corollary 1. *The necessary and sufficient condition for any given basic solution* $x_B = B^{-1}b$ *to be non-degenerate is the linear independence of b and every (m – 1) columns from B.*

Proof:

The condition is necessary. We know that if a solution is non-degenerate ($x_i \neq 0, j = 1, 2, \ldots, m$), b can replace any column of B and still a basis is maintained.

Hence b and any (m – 1) columns of B are L.I.

The condition is sufficient. Let b and any (m – 1) columns of B be L.I. Any column of B can be replaced by b and still a basis is maintained. Hence $x_1, \ldots, x_m$ are all non-zero.

Basis Feasible Solution. *A.B.F.S. is a basic solution which also satisfies the non-negativity restrictions.*

Thus in B.F.S. all basic variables are ≠ 0 *i.e.,* all basic variables are non-negative.

Optimum Basic Feasible Solution. *A basic feasible solution is said to be optimum, if it also optimizes (max. or min.) the objective function.*

Example1:

Show that the feasible solution $x_1 = 1, x_2 = 0, x_3 = 1$ and $Z = 6$ to the system of equations

$$x_1 + x_2 + x_3 = 2$$

$$x_1 - x_2 + x_3 = 2,$$

$$x_i \geq 0$$

which minimizes $Z = 2x_1 + 3x_2 + 4x_3$ is not basic.

Solution:

In the given F.S. there are only two non-zero variables namely x_1 and x_3. If vectors associated to these variables be a1 and a_3 then the determinant of column vectors corresponding to variables x_1, x_3 is $\begin{vmatrix} 1 & 1 \\ 1 & 1 \end{vmatrix}$ = 1 – 1 = 0.

∴ the vectors α_1, α_3 are L.D.

Hence the solution is not a basic solution.

Example 2:

Find all the basic solution of the following system:

$$x_1 + 2x_2 + x_3 = 4$$

$$2x_1 + x_2 + 5x_3 = 5$$

and prove that they are non-degenerate.

Solution:

The given system of equations can be expressed in matrix form as $Ax = b$ where $A = (\alpha_1, \alpha_2, \alpha_3)$

$$\alpha_1 = \begin{bmatrix}1\\2\end{bmatrix}, \alpha_2 = \begin{bmatrix}2\\1\end{bmatrix},$$

$$\alpha_3 = \begin{bmatrix}1\\5\end{bmatrix}, b = \begin{bmatrix}4\\5\end{bmatrix}, x = \begin{bmatrix}x_1\\x_2\\x_3\end{bmatrix},$$

Here n = number of variables = 3 and m = number of equation = 2. Hence there can be almost ${}^3C_2 = 3$ basic solution.

Three sets of two vectors out of a_1, a_2, a_3 are as follows:

$$B_1 = [\alpha_1, \alpha_2] = \begin{bmatrix}1 & 2\\2 & 1\end{bmatrix},$$

$$B_2 = [\alpha_1, \alpha_3] = \begin{bmatrix}1 & 1\\2 & 5\end{bmatrix},$$

$$B_3 = [\alpha_2, \alpha_3] = \begin{bmatrix}2 & 1\\1 & 5\end{bmatrix}.$$

We have $|B_1| = -3 \neq 0$. $|B_2| = 3 \neq 0$. $|B_3| = 9 \neq 0$.

If follows that every set of two vectors of A is L.I. Hence all the three basic solutions exist. If x_{B1}, x_{B2}, x_{B3} are the vectors of corresponding basic variables, then

$$x_{B1} = {}_{B1}^{-1}b = -\frac{1}{2}\begin{bmatrix}1 & -2\\-2 & 1\end{bmatrix}\begin{bmatrix}4\\5\end{bmatrix} = \begin{bmatrix}2\\1\end{bmatrix}$$

$$x_{B2} = B_2^{-1}b = \frac{1}{3}\begin{bmatrix}5 & -1\\-2 & 1\end{bmatrix}\begin{bmatrix}4\\5\end{bmatrix} = \begin{bmatrix}5\\-1\end{bmatrix}$$

$$x_{B3} = B_3^{-1}b = \frac{1}{9}\begin{bmatrix}5 & -1\\-1 & 2\end{bmatrix}\begin{bmatrix}4\\5\end{bmatrix} = \begin{bmatrix}5/3\\2/3\end{bmatrix}.$$

To find all the basic solution. In basis B_1, basis vectors are α_1, $\alpha2$.

Σo

$$x_{B1} = \begin{bmatrix} x_1 \\ x_2 \end{bmatrix} = \begin{bmatrix} 2 \\ 1 \end{bmatrix} \Rightarrow x_1 = 2,\ x_2 = 1.$$

These two variables are the basic variables and the remaining variables x3 is non-basic. It will have value zero. Thus the basic solution associated to the basis B_1 is given by (2, 1, 0). Similarly other basic solutions are (5, 0, –1) and (0, 5/3, 2/3).

In all the three basic solution none of the basic variables is zero, hence they are all non-degenerate.

Example 3:

Which of the following vectors $x = [x_1, x_2, x_3, x_4, x_5, x_6, x_7]$ is a B.F.S. of the system

$$x_1 + 2x_2 + x_3 + 3x_4 + x_5 = 9$$

$$2x_1 + x_2 + 3x_4 + x_6 = 9$$

$$-x_1 + x_2 + x_3 + x_7 = 0,$$

$$x_1, x_2, \ldots, x_7 \geq 0.$$

(i) $x_1 = [2, 2, 0, 1, 0, 0, 0]$

(ii) $x_2 = [0, 0, 9, 0, 0, 9, -9]$

(iii) $x_3 = [3, 3, 0, 0, 0, 0, 0]$

(iv) $x_4 = [0, 0, 0, 0, 9, 9, 0]$

(v) $x_5 = [1, 0, 0, 0, 8, 7, 1]$

(vi) $x_6 = [0, 0, 0, 3, 0, 0, 0]$.

Solution:

We have denoted the column vector by symbol [ ].

(i) The vectors associated to non-zero variables are

$$\alpha_1 = [1, 2, -1],$$

$$\alpha_2 = [2, 1, 1],$$

$$\alpha_4 = [3, 3, 0].$$

Obviously there exist scalars 1, 1 and – 1 such that

$$1.\alpha_1 + 1.\alpha_2 - 1.\alpha_4 = [0, 0, 0] = 0.$$

$\therefore$ vectors a_1, a_2, a_4 associated to non-zero variables are L.D.

Hence x is not a B.F.S.

(ii) The vectors associated to non-zero variables are

$\alpha_3 = [1, 0, 1]$, $\alpha_6 = [0, 1, 0]$, $\alpha_7 = [0, 0, 1]$.

$$\text{Now } |\alpha_3\ \alpha_6\ \alpha_7| = \begin{vmatrix} 1 & 0 & 0 \\ 0 & 1 & 0 \\ 1 & 0 & 1 \end{vmatrix} = 1 \neq 0$$

$\therefore$ vectors α_3, α_6, α_7 are L.I.

Hence x_2 is a basic solution but not a basic feasible solution as x_7 is negative.

(iii) This solution contains 5 zero variables. So it may be a B.F.S. Now vectors associated to non-zero variables are

$\alpha_1 = [1, 2, -1]$, $\alpha_2 = [2, 1, 1]$.

Now these vectors together with the vector $\alpha_5 = [1, 0, 0]$ are L.I. as

$$|\alpha_1\ \alpha_2\ \alpha_5| = \begin{vmatrix} 1 & 2 & 1 \\ 2 & 1 & 0 \\ -1 & 1 & 0 \end{vmatrix} = 3 \neq 0.$$

Hence x_3 is B.F.S. with x_1, x_2, x_5 as basic variables.

Since one of the basic variables namely x_5 vanishes, so this is a degenerate B.F.S.

(iv) This solution contains 5 zero variables. So it may be a B.F.S. Now vectors associated to non-zero variables are $a_5 = [1, 0, 0]$, $a_6 = [0, 1, 0]$. Obviously these vectors together with $a_7 = [0, 0, 1]$ are L.I. Hence x_4 is a B.F.S. with x_5, x_6, x_7 as basic variables. Since one of the basic variables namely x_7 vanishes, so this is degenerate B.F.S.

(v) This solution does not contain alteast $7 - 3 = 4$ variables zero or it contains more than 3 (no. of equations) non-zero variables, so it is not a B.F.S.

(vi) This solution contains one non-zero variable. The vector associated to this variable is $a_4 = [3, 3, 0]$.

$$\text{We have } |\alpha_4\ \alpha_6\ \alpha_7| = \begin{vmatrix} 3 & 0 & 0 \\ 3 & 1 & 0 \\ 0 & 0 & 1 \end{vmatrix} = 3 \neq 0$$

$\therefore$ vectors $\alpha_4, \alpha_6, \alpha_7$ are L.I.

Hence x_6 is a B.F.S. with x_4, x_6, x_7 basic variables. Since basic variables x_6, x_7 also vanish, so it is a degenerate B.F.S.

Example 4:

A manufacture of patent medicines is proposed to prepare a production plan for medicines A and B. There are sufficient ingredients available to make 20, 000 bottle of medicine A and 40,000 bottles of medicine B, but there are only 45,000 bottles into which either of the medicines can be filled. Further, it takes three hours to prepare enough material to fill 1000 bottles of medicine A and one hour to prepare enough material to fill 1000 bottles of medicine B and there are 66 hours available for this operation. The profit is Rs. 8 per bottle for medicine A and Rs. 7 per bottle for medicine B. Formulate this problem as a L.P.P..

Solution:

Suppose the manufacturer produces x_1 thousand bottles of medicine A and x_2 thousand bottles of medicine B.

Since there are only 45,000 bottles available for filling the medicine A and B, therefore we have $x_1 + x_2 \leq 45$.

It takes three hours to prepare enough material to fill 1000 bottles of medicines A. Therefore the time required to fill x_1 thousand bottles of medicine A is $3x_1$ hours.

Similarly, the time required to fill x_2 thousand bottles of medicine B is x_2 hours.

Thus, the total time required to fill x_1 thousand bottles of medicines A and x_2 thousand bottles of medicine B is $(3x_1 + x_2)$ hours.

The total time available for this operation is 66 hours. So, we have $3x_1 + x_2 \leq 66$.

There are sufficient ingredients available to make 20 thousand bottles of medicine A and 40 thousand bottles of medicine B, therefore we have $x_1 \leq 20$, $x_2 \leq 40$.

Since the number of bottles cannot be negative, therefore we have $x_1 \geq 0$, $x_2 \geq 0$.

The profit per bottle for medicine A is Rs. 8 and for medicine B is Rs. 7. So, the total profit in Rs. on x_1 thousand bottles of medicine

A and x_2 thousand bottles of medicines B is given by

$$Z = 8 \times 1000x_1 + 7 \times 1000x_2 = 8000x_1 + 7000x_2.$$

Hence, the linear programming problem formulated for the given problem is

Maximize $Z = 8000x_1 + 7000x_2$,

subject to the constraints:

$$x_1 + x_2 \leq 45$$
$$3x_1 + x_2 \leq 66$$
$$x_1 \leq 20$$
$$x_2 \leq 40$$

and the non-negative restrictions $x_1 \geq 0$, $x_2 \geq\leq 0$.

SYSTEMS OF LINEAR NON-HOMOGENEOUS EQUATIONS

Sometimes we think that we can solve every two simultaneous of the type

$$\left.\begin{aligned} a_1x + b_1y = c_1 \\ a_2x + b_2y = c_2 \end{aligned}\right\}.$$

But it is not so. For example, consider the simultaneous equations

$$\left.\begin{aligned} 3x + 4y = 5 \\ 6x + 8y = 13 \end{aligned}\right\}.$$

There is no set of values of x and y which satisfies both these equations. Such equations are said to be *inconsistent.*

Let us take another example. Consider the simultaneous equations

$$\left.\begin{aligned} 3x + 4y = 5 \\ 6x + 8y = 10 \end{aligned}\right\}.$$

These equations are consistent since there exist values of x and y which satisfy both of these equation. We see that $x = -\frac{4}{3}c + \frac{5}{3}$, $y = c$ constitute a solution of these equation, where c is arbitrary. Thus these equations possess an infinite number of solution.

Now we shall discuss the nature of solution of a system of non-homogeneous linear equations.

Let

$$\left.\begin{array}{l} a_{11}x_1 + a_{12}x_2 + \ldots + a_{1n}x_n = b_1 \\ a_{21}x_1 + a_{22}x_2 + \ldots + a_{2n}x_n = b2 \\ \ldots \ldots \ldots \ldots \ldots \ldots \ldots \ldots \ldots \ldots \\ \ldots \ldots \ldots \ldots \ldots \ldots \ldots \ldots \ldots \ldots \\ a_{m1}x_1 + a_{m2}x_2 + \ldots + a_{mn}x_n = b_m \end{array}\right\} \quad \ldots(1)$$

be a system of m non-homogeneous equations in n unknowns

$x_1, x_2, \ldots, x_n$.

If we write

$$A = \begin{bmatrix} a_{11} & a_{12} & \ldots & a_{1n} \\ a_{21} & a_{22} & \ldots & a_{2n} \\ \ldots & \ldots & \ldots & \ldots \\ a_{m1} & a_{m2} & \ldots & a_{mn} \end{bmatrix}_{m\times n}, \quad X = \begin{bmatrix} x_1 \\ x_2 \\ \ldots \\ x_n \end{bmatrix}_{n\times 1}$$

$$B = \begin{bmatrix} b_1 \\ b_2 \\ \ldots \\ b_m \end{bmatrix}_{m\times 1}$$

where A, X, B are m × n, n × 1 and m × 1 matrices respectively, the above equations can be written in the form of a single matrix equation AX = B.

Any set of values of $x_1, x_2, \ldots x_n$ which simultaneously satisfy all these equations is called a solution of the system (1). When the system of equations has one or more solutions, the equations are said to be ***consistent****, otherwise they are said to be inconsistent.*

The matrix

$$[A\ B] = \begin{bmatrix} a_{11} & a_{12} & \ldots & a_{1n} & b_1 \\ a_{21} & a_{22} & \ldots & a_{2n} & b_2 \\ \ldots & \ldots & \ldots & \ldots & \ldots \\ a_{m1} & a_{m2} & \ldots & a_{mn} & b_m \end{bmatrix}$$

is called the augmented matrix of the given system of equations.

SOME IMPORTANT DEFINITIONS AND RESULTS

Solution of a LPP. *A set of values of the variables x1, x2, ... ,n satisfying the constraints of a LPP is called a solution of the LPP.,*

Feasible Solution of a LPP. A set of values of the variables x_1, x_2, ... ,x_n satisfying the constraints and non-negative restrictions of a LPP is called feasible solution of the LPP.

Optimal Solution of a LPP. A feasible solution of a LPP is said to be optimal (or optimum) if it also optimizes (i.e., maximizes or minimizes as the case may be) the objective function of the problem.

Graphical Solution of a LPP. The solution of a LPP obtained by graphical method i.e., by drawing the graphs corresponding to the constraints and the non-negative restrictions is called the graphical solution of a LPP.

Unbounded Solution. If the value of objective function can be increased or deceased indefinitely, such solutions are called unbounded solutions.

Fundamental Extreme Point Theorem. An optimum solution of a LPP, if it exists, occurs at one of the extreme points (*i.e.,* corner points) of the convex of the set of all feasible solutions.

SOLUTION OF SIMULTANEOUS LINEAR INEQUATIONS

The graph of the solution set of a system of simultaneous inquisitions is the region containing the point (x, y) which satisfy all the inquisitions of given system simultaneously.

To draw the graph of the simultaneous linear inquisitions *i.e.,* to find the solution set of the simultaneous linear inquisitions, we find the region of the xy-plane, common to all the portions comprising the solution sets of the given inquisitions. If there is no region common to all the solutions of the given inquisitions, we say that the solution set of the system of inquisition is empty. It should be noted that the solution set of simultaneous linear inquisition may be an empty set or it may be the region bounded by the straight lines corresponding to given linear inquisitions or it may be an unbounded region with straight line boundaries.

FORMULATION AS A LINEAR PROGRAMMING PROBLEMS

Step 1. We study the situation to find the day decision to be made and in this connection loading for variables helps considerable.

Step 2. Select symbols for variable quantities identified in step I.

Let the number units of the products A, B and C manufactured daily be designated by x_1, x_2 and x_3 respectively.

Step 3. Express feasible alternatives mathematical in terms of the variables. These feasible alternatives are those which are physically, economically and financially possible. Since it is not possible to manufacture any negative quantities, it is quite obvious that in the present situation feasible alternatives are sets of values of x_1, x_2 and x_3 satisfying $x_1 \geq 0$, $x_2 \geq 0$, $x_3 \geq 0$.

Step 4. Identify the objective quantitatively and express it as a linear function of variables. The objective here is to maximize the profit. In view of the assumption that all the units produced are consumed in the market, it is given by the linear function

$$z = 3x_1 + 2x_2 + 4x_3.$$

Step 5. *Express in words the* **influencing factors** *or* **constraints** (*or* **restrictions**) *which occur generally because of the constraints on availability (resources) or requirements (demands). Express these restrictions also as linear equalities/inequalities in terms of the variables.* Here in order to produce x_1 units of product A, x_2 units of product B and x_3 units of product C, the total times needed on machines P and Q are given by

$$4x_1 + 3x_2 + 5x_3 \text{ and } 2x_1 + 2x_2 + 4x_3$$

respectively. Since the manufacturer does not have more than 2,000 minutes available on machine P and 2,500 minutes available on machine Q, we must have

$$4x_1 + 3x_2 + 5x_3 \leq 2{,}000$$

and $$2x_1 + 2x_2 + 4x_3 \leq 2{,}000.$$

Also, the manufacturer has to satisfy the following given additional restrictions:

$$x_1 \leq 150,\ x_1 \geq 100;$$

$$x_2 \geq 200 \text{ and } x_3 \geq 50.$$

Hence the manufacturer's problem can be put in the following mathematical form:

Determine three non-negative real numbers x_1, x_2 and x_3 such that

(i) $4x_1 + 3x_2 + 5x_3 \leq 2{,}000$

(ii) $2x_1 + 2x_2 + 4x_3 \leq 2{,}500$

(iii) $100 \leq x \leq 150$, $x_2 \leq 250$, $x_3 \leq 50$

and for which the expression (objective function)

$$z = 3x_1 + 2x_2 + 4x_3$$

may be a maximum (greatest).

This problem is called linear programming problem since a linear function of three variable x_1, x_2 and x_3 defined by

$$z = 3x_1 + 2x_2 + 4x_3$$

$$x_1 + x_2 \leq 24. \qquad ...(4)$$

Also the number of necklaces and bracelets manufactured can never be negative, therefore

$$x_1 \geq 0, x_2 \geq 0.$$

Hence, the linear programming problem formulated from the given problem is as follows:

$$\text{Maximize } Z = 100x_1 + 300x_2$$

subject to the constraints

$$x_1 + 2x_2 \leq 32$$

$$x_1 + x_2 \leq 24$$

and the non-negative restrictions $x_1 \geq 0$, $x_2 \geq 0$.

Example:

A tyre factory produces three types of tyre T_1, T_2, T_3. Three different types of chemicals say C_1, C_2, C_3 are required for production. One T_1 tyre needs 2 units of C_1, 3 units of C_3; one T2 tyre needs 3 units 3 units of C_1, 2 units of C_2 and 2 units of C_3; and one T3 tyre needs 5 units of C_2 and 4 units of C_3. The factory has only a stock of 20 units of C_1, 25 units of C_2 and 30 units of C_3. Further the profit from the sale of one tyre T_1 is Rs. 6, one tyre T_2 is Rs. 10 and of one tyre T_3 is Rs. 8. Assuming that the factory can sell all that it produces, formulate a linear programming problem to maximize its profits.

Solution:

Let the factory produce x_1 tyres of type T_1, x_2 tyres of type T_2 and x_3 tyres of type T_3.

The given information can be put in a tabular form as given below:

	Tyre T_1	*Tyre T_2*	*Tyre T_3*	*Total chemical available*
Chemical C_1	2	3	0	20
Chemical C_2	0	2	5	25
Chemical C_3	3	2	4	30
Profit in Rs. (per tyre)	6	10	8	

The total profit Z in Rs. is given as

$$Z = 6x_1 + 10x_2 + 8x_3. \quad ...(1)$$

The total quantity of chemical C_1 required = $(2x_1 + 3x_2)$ units.

Since the factory has a stock of 20 units of chemical C_1, therefore we have

$$2x_1 + 3x_2 \leq 20 \quad ...(2)$$

Similarly, considering the total quantity of the chemicals C_2 and C_3 required, we have

$$2x_2 + 5x_3 \leq 25 \quad ...(3)$$

and $$3x_1 + 2x_2 + 4x_3 \leq 30. \quad ...(4)$$

Also we have

$$x_1 \geq 0, x_2 \geq 0, x_3 \geq 0, \quad ...(5)$$

since the factory cannot produce negative quantities.

Hence, the linear programming problem formulated from the given problem is as follows:

Maximize $Z = 6x_1 + 10x_2 + 8x_3$

subject to the constraints

Example 2:

The objective of a diet problem is to ascertain the quantities of certain foods that should be eaten to meet certain nutritional requirement at a minimum cost. The consideration is limited to mild, green vegetables and eggs, and to vitamins A, B, C. The number of milligrams of each of these vitamins contained within a unit of each food and their daily minimum requirement along with the cost of each food is given in the table below:

Vitamin	*Litre of Milk*	*Kg. of Vegetables*	*Dozen of Eggs*	*Minimum Daily Requirement*
A	1	1	10	1 mg.
B	100	10	10	50 mg.
C	10	100	10	10 mg.
Cost in Rs.	20	10	8	

Formulate a linear programming problem for this diet problem.

Solution:

Let the daily diet consist of x_1 litres of milk, x_2 kgs. of vegetables and x_3 dozens of eggs.

Then total cost Z per day in Rs. is given by

$$Z = 20x_1 + 10x_2 + 8x_3. \quad \text{...(1)}$$

Total amount of vitamin A in the daily diet is

$$(x_1 + x_2 + 10x_3) \text{ mg.,}$$

which should be at least equal to 1 mg., therefore

$$x_1 + x_2 + 10x_3 \geq 1. \quad \text{...(2)}$$

Similarly, considering the total amounts of vitamins B and C in the daily diet, we have

$$100x_1 + 10x_2 + 10x_3 \geq 50 \quad \text{...(3)}$$

and $$10x_1 + 100x_2 + 10x_3 \geq 10 \quad \text{...(4)}$$

Since the quantities of different food items to be consumed cannot be negative, therefore we have

$$x_1 \geq 0,\ x_2 \geq 0,\ x_3 \geq 0.$$

Hence, the linear programming problem formulated from the given diet problem is:

Minimize $Z = 20x_1 + 10x_2 + 8x_3$,

subject to the constricts

$$x_1 + x_2 + 10x_3 \geq 1$$

$$100x_1 + 10x_2 + 10x_3 \geq 50$$

$$10x_1 + 100x_2 + 10x_3 \geq 10$$

and the non-negative restrictions

$$x_1 \geq 0, x_2 \geq 0, x_3 \geq 0.$$

Example 3:

A certain farming organization operates three farms of comparable productivity. The output each farm is limited both by the usable acreage and by the amount of water available for irrigation. The data for the up coming season are the following:

Farm	*Usable acreage*	*Water available in acre metre*
1	*400*	*150*
2	*600*	*200*
3	*300*	*90*

The organization is considering three crops for planting which differ primarily in their expected profit per acre and in their consumption of water. Furthermore, the total acreage that can be devoted to each of the crops is limited by the amount of appropriate harvesting equipment available:

Crop	*Maximum*	*Water consumption acreage*	*Expected profit in acre metre per acre*
A	*700*	*0'5*	*Rs. 4,000*
B	*800*	*0'4*	*Rs. 3,000*
C	*300*	*0'3*	*Rs. 1,000*

In order to maintain a uniform workload among the farms, it is the policy of the organization that the percentage of the usable acreage planted must be the same at each farm. However, any combination of the crops may be grown at any of the farms. The organization wishes to know how much of each crop should be planted at the respective farms in order to maximize expected profit. Formulate this as a linear programming problem.

Formulation as a Linear Programming Problem

Step 1. The key decision to be made is to determine the number of acres to be devoted to the three crops at each of the three farms.

Step 2. Let x_j j = 1, 2, ... ,9, represent these nine quantities identified in Step 1. Let us demonstrate these variables as follows:

	(Allocation in acres)		
		crop	*Farm*
	1	2	3
A	x_1	x_2	x_3
B	x_4	x_5	x_6
C	x_7	x_8	x_9

Step 3. Feasible alternatives are the sets of values of x_1, x_2 and x_3 satisfying $x_1, x_2, x_3 \geq 0$, since negative acreages are meaningless.

Step 4. The objective is to maximize the total net profit (in rupees) of the organization, *i.e.,* to maximize

$z = 4000\ (x_1 + x_2 + x_3) + 3000(x_4 + x_5 + x_6) + 1000(x_7 + x_8 + x_9)$.

Step 5. Constraints are the availability of usable land, water for irrigation, requirement of minimum acreage and social equality of percentage of the usable acreage planted for each farm. These are

1 Land: $x_1 + x_4 + x_7 \leq 400$

$x_2 + x_5 + x_8 \leq 600$

$x_3 + x_6 + x_9 \leq 300$

2 Water: $5x_1 + 4x_4 + 3x_7 \leq 1500$

$5x_2 + 4x_5 + 3x_8 \leq 2000$

$5x_3 + 4x_6 + 3x_9 \leq 900$

3 Crop requirement:

$x_1 + x_2 + x_3 \leq 700$

$x_4 + x_5 + x_6 \leq 800$

$x_7 + x_8 + x_9 \leq 300$

4 *Social equality:*

$$\frac{x_1 + x_4 + x_7}{400} = \frac{x_2 + x_5 + x_8}{600}$$

$$\frac{x_2 + x_5 + x_8}{600} = \frac{x_3 + x_6 + x_9}{300}$$

$$\frac{x_1 + x_2 + x_3}{3} = \frac{x_1 + x_4 + x_7}{400}$$

i.e., $3(x_1 + x_4 + x_7) - 2(x_2 + x_5 + x_8) = 0$

$x_2 + x_5 + x_8 - 2(x_3 + x_6 + x_9) = 0$

$4(x_3 + x_6 + x_9) - 3(x_1 + x_4 + x_7) = 0$

This complete the formulation.

$2x_1 + 3x_2 \le 20$

$2x_2 + 5x_3 \le 25$

$3x_1 + 2x_2 + 4x_3 \le 30$

and the non-negative restrictions

$x_1 \ge 0, x_2 \ge 0, x_3 \ge 0.$

Example 4:

A complete unit of a certain product consists of four units of component A and three units of component B. The two components (A and B) are manufactured from two different raw materials of which 100 units and 200 units respectively are available. Three departments are engaged in the production process wit each department using a different method for manufacturing the components.

The following table gives the raw material requirements per production run and the resulting of each component. The objective is to determine the number of production runs for each department which will maximize the total number of component units of the final product:

Department	*Input per run (units) Raw material*		*Output per run (units) Component*	
	I	II	A	B
A	7	5	6	4
B	4	8	5	8
C	2			
		7		
			7	
				3

Solution:

Let x_1, x_2, x_3 be the number of production runs for the departments A, B, C respectively.

The raw materials available give the restrictions

$$7x_1 + 4x_2 + 2x_3 \leq 100 \quad ...(1)$$

$$5x_1 + 8x_2 + 7x_3 \leq 200 \quad ...(2)$$

The total numbers of units produced by three departments:

$6x_1 + 5x_2 + 7x_3$ (component A) and $dx_1 + 8x_2 + 3x_3$ (component B).

Now final product requires 4 units of component A and 3 units of component B. So the maximum number of units of the final product cannot exceed the smaller value of

$$\frac{1}{4}(6x_1 + 5x_2 + 7x_3) \text{ and } \frac{1}{3}(4x_1 + 8x_2 + 3x_3).$$

Thus the objective function becomes

$$\text{Max. } Z = \sin\left[\frac{1}{4}(6x_1 + 5x_2 + 7x_3), \frac{1}{3}(4x_1 + 8x_2 + 3x_3)\right].$$

Since the objective function is not linear, a suitable transformation can be used to convert this into a L.P.P.

$$\text{Let, min.}\left[\frac{1}{4}(6x_1 + 5x_2 + 7x_3), \frac{1}{3}(4x_1 + 8x_2 + 3x_3)\right] = v.$$

$$\text{Then } \frac{1}{4}(6x_1 + 5x_2 + 7x_3) \geq v \text{ and } \frac{1}{3}(4x_1 + 8x_{12} + 3x_3) \geq v.$$

Thus we get the L.P.P. as follow:

Maximize $\quad Z = v.$

subject to the constraints:

$$6x_1 + 5x_2 + 7x_3 - 4v \geq 0$$

$$4x_1 + 8x_2 + 3x_3 - 3v \geq 0$$

$$7x_1 + 4x_2 + 2x_3 \leq 100$$

$$5x_1 + 8x_2 + 7x_3 \leq 200$$

and the non-negativity restrictions $x_1, x_2, x_3, v \geq 0$.

Example 5(a):

The owner of the Metro Sports wishes to determine how many advertisements to place in the selected three monthly magazines A, B and C. His objective is to advertise in such a way that total exposure to principal buyers of expensive sports good is maximized. Percentages of readers for each magazine are known. Exposure in any particular magazine is the number of advertisements placed multiplied by the number of principal buyers. The following data may be used:

Exposure Category	*Magazines*		
	A	B	C
Readers (in Lacks)	1	0.6	0.4
Principal Buyers	10%	15%	7%
Cost per advertisement (in Rs.)	5000	4500	4250

The budgeted amount is at most Rs. 1 take for the advertisements. The owner has already decided that magazine A should have no more than 6 advertisements and that B and C, each have at last two advertisements, formulate a LP model for the problem.

Solution:

Let x_1, x_2, x_3 be the number of insertions in magazines A, B and C respectively.

Then the LP formulation will be as follows:

$$\text{Maximize } Z = (10\% \text{ of } 100000)\, x_1 + (15\% \text{ of } 60000)\, x_2 + (7\% \text{ of } 40000)\, x_3$$

or $$\text{Maximize } Z = 10000x_1 + 90000x_2 + 2800x_3$$

subject to the constitutive restrictions

$$x_1, x_2, x_3 \geq 0.$$

Example 5(b):

A manufacturer of biscuits is considering four types of gift packs containing three types of buckets: orange cream (OC), chocolate cream (CC), and wafers (W). Market research conducted recently to assess the preferences of the consumers shows the following types of assortments to be in good demand:

Assortments	*Contents*	*Selling price per kg (Rs.)*
A	*Not less than 40% of OC*	20
	Not more than 20% of CC	
	Any quantity of W	
B	*Not less than 20% of OC*	25
	Not more than 40% of CC	
	Any quantity of W	
C	*Not lass than 50% of OC*	
		22
	Not more than 10% of CC	
	Any quantity of W	
D	*Not restrictions*	12

For the biscuits, the manufacturing capacity and costs are given below:

Biscuits variety	*OC*	*CC*	*W*
Plant capacity (kg/day)	200	200	150
Manufacturing cost (Rs/kg.)	8	9	7

Formulate a linear programming model to find the production schedule which maximizes the profit assuming that there are no market restrictions.

Solution:

Let x_{ij} kg. be the quantity used in ith assortment of the jth type of biscuits, where i = A, B, C, D and j = OC, CC, W (number 1, 2, 3 respectively).

Then in assortment A, x_{A1}, x_{A2}, x_{A3} denote the quantity in kg. of OC, CC and W type of biscuits respectively. Similarly we can interpret for other assortments.

Now the given data can be put in the form of L.P.P. as follows:

Maximize $Z = 20\ (x_{A1} + x_{A2} + x_{A3}) + 25\ (x_{B1} + x_{B2} + x_{B3}) + 22\ (x_{C1} + x_{C2} + x_{C3}) + 12\ (x_{D1} + x_{D2} + x_{D3}) - 8\ (x_{A1} + x_{B1} + x_{C1} + x_{D1}) -$

or Maximize $Z = 12x_{A1} + 11x_{A2} + 13x_{A3} + 17x_{B1} + 16x_{B2} + 18x_{B3} + 14x_{C1} + 13x_{C2} + 15x_{C3} + 4x_{D1} + 3x_{D2} + 5x_{D3}$

subject to the constraints:

$x_{A1} \geq 0.40\ (x_{A1} + x_{A2} + x_{A3})$

$x_{A2} \leq 0.20\ (x_{A1} + x_{A2} + x_{A3})$ gift pack A

$x_{B1} \geq 0.20\ (x_{B1} + x_{B2} + x_{B3})$

$x_{B2} \leq 0.40\ (x_{B1} + x_{B2} + x_{B3})$ gift pack B

$x_{C1} \geq 0.50\ (x_{C1} + x_{C2} + x_{C3})$

$x_{C2} \leq 0.10\ (x_{C1} + x_{C2} + x_{C3})$ gift pack C

Plant capacity constraints are: $x_{A1} + x_{B1} + x_{C1} + x_{D1} \leq 200$

$x_{A2} + x_{B2} + x_{C2} + x_{D2} \leq 200$

$x_{A3} + x_{B3} + x_{C3} + x_{D3} \leq 150$

and non-negativity restrictions are $x_{ij} \geq 0$ (i = A, B, C, D; j = 1, 2, 3).

Example 5(c):

A company has two grades of inspectors, I and II, who are to be assigned for a quality control inspection. It is required that at least 2000 pieces be inspected per 8 hour day. Grade I inspectors can check pieces at the rate of 50% hour with the accuracy of 97%. Grade II inspectors can check pieces at the rate of 40% hour with the accuracy of 95%. The wage rate of Grade I inspector is Rs. 4.50% hour and that of Grade II is Rs. 2.50/hour. Each time an error is made by an inspector, the cost to the company is one rupee. The company has available for the inspection job, 10 grade I and 5 grade II inspectors. Formulate the problem to minimize the total cost of inspection.

Solution:

Let the no. of inspectors of grade I and II be x_1 and x_2 respectively. They will inspect (8×50) x1 + $(8 \times 40)\ x_2$ pieces daily. As the company requires attest 2000 items to be inspected daily, of the constraint is

$$(8 \times 50)\ x_1 + (8 \times 40)\ x_2 \geq 2000 \text{ i.e., } 5x_1 + 4x_2 \geq 25. \quad ...(1)$$

The limitations on the available numbers of inspectors give the constraints

$$x_1 \leq 10,\ 2 \leq 5 \qquad \ldots(2)$$

Further the company is bearing two types of costs, the wages of inspectors and the costs of inspection errors.

The cost of each inspector per hour is as follows:

$$\text{Grade I : Rs. } \left[4.50 + 1 \times \left(\frac{3}{100} \times 50\right)\right] = \text{Rs. } 6.00/\text{hour}$$

$$\text{Grade II : Rs. } \left[2.50 + 1 \times \left(\frac{5}{100} \times 40\right)\right] = \text{Rs. } 4.50/\text{hour.}$$

Now the objective of the company is to minimize the total cost of inspection at the above rates for each day.

Thus the objective function is

$$\text{Minimize } Z = 8 \times 6.00 \times x_1 + 8 \times 4.\ 50 \times x_2 = 48x_1 + 36x_2 \qquad \ldots(3)$$

Hence the required L.P.P. is

$$\text{Minimize } Z = 48x_1 + 36x_2$$

subject to the constraints: $5x_1 + 4x_2 \geq 25$

$x_1 \leq 10,\ x_2 \leq 5$

and the non-negativity restrictions $x_1, x_2 \geq 0$.

Example 6:

According to the medical experts it is necessary for an adult to consume at least 75 gms of proteins, 85 gms of fats and 300 gms of carbohydrates daily. The following table gives the analysis of the food items readily available in he market with their respective costs.

Food	*Food value (in gm.) per 100 gms*			*Cost in*
Type	*Proteins*	*Fats*	*Carbohydrates*	*Rs. per kg*
A	18.0	15.0	–	3.0
B	16.0	4.0	7.0	4.0
C	4.0	20.0	2.5	2.0
D	5.0	8.0	40.0	1.5
Minimum daily	75.0	85.0	300.0	

Formulate a linear programming problem for an optimum diet.

Solution:

Let the daily diet consist of x_1 kg of food A, x_2 kg of food B, x_3 kg of food C and x_4 kg of food D.

Then the total cost per day in Rs. is

$$Z = 3x_1 + 4x_2 + 2x_3 + 1.5x_4. \qquad ...(1)$$

Total amount of proteins in the daily diet is

$$(180x_1 + 160x_2 + 40x_3 + 50x_4).$$

Since the minimum daily requirement of proteins is 75 gms, therefore we have

$$180x_1 + 160x_2 + 40x_3 + 50x_4 \geq 75. \qquad ...(2)$$

Similarly, considering the total amounts of fats and carbohydrates in the diet, we have

$$150x_1 + 40x_2 + 200x_3 + 80x_4 \geq 85 \qquad ...(3)$$

and $$70x_2 + 25x_3 + 400x_4 \geq 300. \qquad ...(4)$$

Since, the daily diet cannot contain quantities with negative values of any food item, therefore

$$x_1 \geq 0, x_2 \geq 0, x_3 \geq 0, x_4 \geq 0. \qquad ...(5)$$

Hence, the linear programming problem formulated for the given diet problem is:

Minimize $Z = 3x_1 + 4x_2 + 2x_3 + 1.5x_4$

subject to the constraints

$$18x_1 + 160x_2 + 40x_3 + 50x_4 \geq 75$$

$$150x_1 + 40x_2 + 200x_3 + 80x_4 \geq 85$$

$$70x_2 + 25x_3 + 400x_4 \geq 300$$

and the non-negative restrictions

$$x_1 \geq 0, x_2 \geq 0, x_3 \geq 0 \text{ and } x_4 \geq 0.$$

Example 7:

A manufacture of patent medicines is proposed to prepare a production plan for medicines A and B. There are sufficient ingredients available to make 20, 000 bottle of medicine A and 40,000 bottles of medicine B, but there are only 45,000 bottles into which either of the medicines can be filled. Further, it takes three hours to prepare enough material to fill 1000 bottles of medicine A and one hour to prepare enough material to fill 1000 bottles of medicine B and there are 66 hours available for this operation. The profit is Rs. 8 per bottle for medicine A and Rs. 7 per bottle for medicine B. Formulate this problem as a L.P.P.

Solution:

Suppose the manufacturer produces x_1 thousand bottles of medicine A and x_2 thousand bottles of medicine B.

Since there are only 45,000 bottles available for filling the medicine A and B, therefore we have $x_1 + x_2 \leq 45$.

It takes three hours to prepare enough material to fill 1000 bottles of medicines A. Therefore the time required to fill x_1 thousand bottles of medicine A is $3x_1$ hours.

Similarly, the time required to fill x_2 thousand bottles of medicine B is x_2 hours.

Thus, the total time required to fill x_1 thousand bottles of medicines A and x_2 thousand bottles of medicine B is $(3x_1 + x_2)$ hours.

The total time available for this operation is 66 hours. So, we have $3x_1 + x_2 \leq 66$.

There are sufficient ingredients available to make 20 thousand bottles of medicine A and 40 thousand bottles of medicine B, therefore we have $x_1 \leq 20$, $x_2 \leq 40$.

Since the number of bottles cannot be negative, therefore we have $x_1 \geq 0$, $x_2 \geq 0$.

The profit per bottle for medicine A is Rs. 8 and for medicine B is Rs. 7. So, the total profit in Rs. on x_1 thousand bottles of medicine A and x_2 thousand bottles of medicines B is given by

$$Z = 8 \times 1000x_1 + 7 \times 1000x_2 = 8000x_1 + 7000x_2.$$

Hence, the linear programming problem formulated for the given problem is

Maximize $Z = 8000x_1 + 7000x_2$,

subject to the constraints:

$x_1 + x_2 \leq 45$

$$3x_1 + x_2 \leq 66$$
$$x_1 \leq 20$$
$$x_2 \leq 40$$

and the non-negative restrictions $x_1 \geq 0$, $x_2 \geq\leq 0$.

Example 8:

A firm manufactures two types of products A and B and sells them at a profit of Rs. 2 on type A and Rs. 3 on type B. Each product is processed on two machines E and F. Type A requires one minute of processing time on E and two minutes on F, type B requires one minute on E and one minute on F. The machine E is available for not more then 6 hours 40 minutes while machine F is available for 10 hours during any working day. Formulate the problem as a linear programming problem.

Solution:

Let the manufacturer produce x_1 units of the product of type A and x_2 units of the product of type B.

The given information can be systematically arranged in the form of the following table:

Machine	*Time of processing (minutes per unit)*		*Time available*
	Type A (x_1 units)	*Type B (x_2 units)*	*(minutes)*
E	1	1	400
F	2	1	600
Profit per unit	Rs. 2		Rs. 3

Since the machine E takes 1 minute time for processing a unit of type A and 1 minute time of processing a unit of type B, therefore the total time required on machine E is $(x_1 + x_2)$ minutes.

But the machine E is available for not more than 6 hours and 40 minutes = 400 minutes, therefore we have

$$x_1 + x_2 \leq 400.$$

Similarly, the total time (in minutes) required on machine F is $2x_1+x_2$.

Since the machine F is available for not more than 10 hours = 600 minutes, therefore we have

$$2x_1 + x_2 \leq 600.$$

Since it is not possible to produce negative quantities, so we have

$$x_1 \geq 0,\ x_2 \geq 0.$$

The profit on type A is Rs. 2 per unit so the profit on selling x_1 units of type A will be Rs. $2x_1$. Similarly, the profit on selling x_2 units of type B will be Rs. $3x_2$.

Therefore, the total profit (in Rs.) on selling x_1 units of type A and x_2 units of type B is

$$Z = 2x_1 + 3x_2.$$

Hence the required linear programming problem formulated for the given problem is

$$\text{Maximize } Z = 2x_1 + 3x_2$$

subject to the conditions

$$x_1 + x_2 \leq 400$$

$$2x_1 + x_2 \leq 600$$

$$x_1 \geq 0,\ x_2 \geq 0.$$

Example 9:

A manufacturer produces three models (I, II and III) of a certain product. He uses two types of raw materials (A and B) of which 4000 and 6000 units respectively are available. The raw material requirements per unit of the three models are given below:

Raw Material	*Requirement per unit of given model*		
	I	II	II
A	2	3	5
B	4	2	7

The labour time for each unit of model I is twice that of model II and three times that of model. III. The entire labour force of the factory can produce the equivalent of 2500 units of model I. A market survey indicates that the minimum demand of the three models are 500, 500 and 375 units respectively. However, the ratio of the number of units produced must be equal to 3:2:5. Assume that the profits per unit of models I, II and III are rupees 60, 40 and 100 respectively. Formulate the problem as a L.P.P. in order to determine the number of units of each product which will maximize profit.

Solution:

Let the manufacturer produce x_1, x_2, x_3 units of models I, II and III respectively. Since the profit per unit on model I, II and III is Rs. 40 and Rs. 100 respectively, the objective function is to maximize the profit:

$$Z = 60x_1 + 40x_2 + 100x_3 \qquad ...(1)$$

The raw materials used and available give rise to the following constraints respectively:

$$2x_1 + 3x_2 + 5x_3 \text{ £ } 4000 \qquad ...(2)$$

$$4x_1 + 2x_2 + 7x_3 \leq 6000 \qquad ...(3)$$

Now if t be the labour time required for one unit of model I, then the time required for one unit of model II will be $\frac{1}{2}t$ and that for the model III will be $\frac{1}{3}t$. As the factory can produce 2500 units of model I, so the restriction on the production time will be

$$tx_1 + \frac{1}{2}tx_2 + \frac{1}{3}tx_3 \leq 25000t \text{ } i.e., \text{ } x_1 + \frac{1}{2}x_2 + \frac{1}{3}x_3 \leq 2500. \qquad ...(4)$$

Now according to the market demand, we must have

$$x_1 \leq 500,\ x_2 \leq 500,\ x_3 \leq 375. \qquad ...(5)$$

Further the ratio of the number of different types of models is 3:2:5

i.e., $x_1 = 3k,\ x_2 = 2k,\ x_3 = 5k.$

These give rise to the constraint

$$\frac{1}{3}x_1 = \frac{1}{2}x_2 \text{ and } \frac{1}{2}x_2 = \frac{1}{5}x_3 \qquad ...(6)$$

Thus, the linear programming problem is as follows:

Maximize $Z = 60x_1 + 40x_2 + 100x_3$

subject of the conditions:

$$2x_1 + 3x_2 + 5x_3 \leq 4000$$

$$4x_1 + 2x_2 + 7x_3 \leq 6000$$

$$x_1 + \frac{1}{2}x_2 + \frac{1}{2}x_2 \leq 2500$$

$$\frac{1}{3}x_1 = \frac{1}{2}x_2, \frac{1}{2}x_2 = \frac{1}{5}x_3$$

and $x_1 \geq 500, x_2 \geq 500, x_3 \geq 375.$

EXERCISES

1. A toy company manufactures two types of dolls; an ordinary doll A and a deluxe doll B. Each type of doll B takes twice as long to produce as one of type A. It is given that the company would have time to make a maximum of 2000 dolls per day if it produces only the ordinary version. The supply of plastic is sufficient to produce 1500 dolls per day (both A and B combined). The deluxe version requires a fancy dress of which there are only 600 pieces per day available. If the company makes profit of Rs. 5 per doll respectively on doll A and doll B, formulate the problem as a linear programming problem to maximize the profit.

2. A furniture firm manufactures chairs and tables, each requiring the use of three machines A, B and C. Production of one chair requires 2 hours on machine A, 1 hour on machine B and 1 hour on machine C. Each table requires 1 hour each on machines A and B and 3 hours on machine C. The profit realized by selling one chair is Rs. 30 while for a table the figure is Rs. 60. The total time available per week on machine A is 70 hours, on machine B is 40 hours, and on machine C is 90 hours. How many chairs and tables should be made per wheel so as to maximize the profit? Formulate a mathematical model for the problem.

3. A diet is to contain at lest 4000 units of carbohydrates, 500 units of fat and 300 units of protein. Two foods A and B are available. Food A costs Rs. 2 per units and food B costs Rs. 4 per unit. A unit of food A contains 10 units of carbohydrates, 20 units of fat and 15 units of protein. A unit of food B contains 25 units of carbohydrates, 10 units of fat and 20 units of protein. Formulate the problem as a LPP so as to find the minimized cost for a diet that consists of a mixture of these two foods and also meets the minimum nutrition requirements.

4. A company manufactures two kinds of leather purses, A and B. A is a high quality purse and B is lower quality. The sales of each of these purses A and B can profit of Rs. 4 and Rs. 3 respectively. Each purse of type A requires twice as much time a purse of type B, and if all purses are of type B, the company could make 1000 purses per day. The supply of leather is sufficient for only 800 purses per day (both A and B combined). Purse A requires a fancy buckle, and only 400 buckles per day are available. There are only 700 buckles available for purse B. What should be the daily production of each type of purse to get the maximum profit? Formulate the problem as a LPP.

5. A firm can produce three types of cloth, say: A, B and C. Three kinds of wool are required for it, say red wool, green wool and blue wool. Unit length of type A cloth needs 2 yards of red wool and 3 green wool; one unit length of type B cloth needs 3 yards of red wool, 2 yards of green wool and 2 yards of blue wool and one unit length of type C cloth needs 5 yards of green wool and 2 yards of blue wool. The firm has only a stock of 8 yards of red wool, 10 yards of green wool and 15 yards of blue wool. It is assumed that the income obtained from one unit length of type A cloth is Rs. 3.00, of type B cloth is Rs. 5.00 and of type C cloth is Rs. 4.00 Formulate this problem as a linear programming model to maximize the income from the finished cloth.

6. In a chemical industry two products A and B are made involving two operations. The production of B also results in a by-product C. The product A can be sold at Rs. 3 profit per unit and B at Rs. 8 profit per unit. The by-product C has a profit of Rs. 2 per unit, but it cannot be sold as the destruction cost is Rs. 1 per unit. Forecasts show that upto 5 units of C can be sold. The company gets 3 units of C for each units of A and B produced. Forecasts whose that they can sell all the units of A and B produced. The manufacturing times are 3 hours per unit for A on operation one and two receptively. Because the product C results from producing B, no time is used in producing C. The available times are 18 and 21 hours of operation one and two respectively. The company questions: how mush A and B should be produced keeping C in mind to make the highest profit P. Formulate LP model for this problem.

7. A resourceful home decorator manufactures two types of lamps say A and B. Both lamps go through two technicians, first a cutter,

second a finisher. Lamp A requires 2 hours of the cutter's time and 1 hour of the finisher's time. Lamp B requires 1 hour of cutter's and 2 hours of finisher's time. The cutter has 104 hours and finisher has 76 hours of time available each month. Profit on the lamp A is Rs. 6 and on the lamp B is Rs. 11. Assuming that he can sell all that he produces, how many of each type of lamps should he manufacture per month to obtain the best returns? Formulate a LPP for this problem.

8. A factory produces two products A and B. To manufacture one unit of product A. A machine has to work 1½ hours and a craftsman has to work for 2 hours. To manufacture one unit of product B, the machine has to work 3 hours and the craftsman for one hour. In a week the factory can avail of 80 hours of machine time and 70 hours of craftsman's time. The profit on the sale of each unit of A and B is of Rs. 10 and Rs. 8 respectively. If the manufacturer can sell all the items produced, how many of each should he produced to get the maximum profit per week?

   Formulate the problem as a LPP.

9. A farmer has 100 acre farm. He can sell all tomatoes, lettuce or radishes he can raise. The price he can obtain is Rs. 1.00 per kg for tomatoes, Rs. 0.75 a head for lettuce and Rs. 2.00 per kg for radishes. The average yield per-acre is 2,000 kg of tomatoes, 3000 heads of lettuce and 1000 kg of radishes. Fertilizer is available at Rs. 0.50 per kg and the amount required for sowing, cultivating and harvesting per acre is 5 man-days for tomatoes and radishes, and 6 man-days for lettuce. A total of 400 man-days of labor are available at Rs. 20.00 per man-day. Formulate this problem as a linear programming model to maximize the farmer's total profit.

Solve graphically the following linear programming problems:

10. Max. $Z = x_1 + 2$

    s.t. $x_1 + 2x_2 \leq 2000,$

    $x_1 + x_2 \leq$

    $x_2 \leq 600$

    and $x_1, x_2 \geq 0.$

11. Max. $Z = 8x_1 + 7x_2$

    s.t. $3x_1 + x_2 \leq 66,000$

$x_1 + x_2 \le 45{,}000$

$x_1 \le 20{,}000 \; x_2 \le 40{,}000$

and $x_1, x_2 \ge 0.$

12. Min. $Z = 1.5x_1 + 2.5x_2$

s.t. $x_1 + 3x_2 \ge 3,$

$x_1 + x_2 \ge 2$

and $x_1, x_2 \ge 0.$

13. Max. $Z = 6x_1 + 11x_2$

s.t. $2x_1 + x_2 \le 104,$

$x_1 + 2x_2 \le 76$

and $x_1, x_2 \ge 0.$

14. Max. $Z = 5x_1 + 7x_2$

s.t. $x_1 + x_2 \le 4,\; 3x_1 + 8x_2 \le 24,\; 10x_1 + 7x_2 \le 35$

and $x_1, x_2 \ge 0.$

15. Min. $Z = 3x_1 + 5x_2$

s.t. $-3x_1 + x_2 \le 12,\; 2x_1 - x_2 \ge -2,\; 2x_1 + 3x_2 \ge 12,\; x_1 \le 4,\; x_2 \ge 2$

and $x_1, x_2 \ge 0.$

16. Max. $Z = 3x_1 + 2x_2$

s.t. $x_1 + x_2 \le 4,\; x_1 - x_2 \le 2,\; x_1, x_2 \ge 0.$

17. Max. $Z = 3x_1 + 4x_2$

s.t. $x_1 - x_2 \le -1,\; -x_2 + x_2 \le 0,\; x_1, x_2 \ge 0.$

18. Min. $Z = x_1 + x_2,$

s.t. $5x_1 + 10x_2 \le 5,\; x_1 + x_2 \ge 1,\; x_2 \le 4,\; x_1, x_2 \ge 0.$

19. Max. $Z = 0.75x_1 + x_2$

s.t. $x_1 - x_2 \ge 0,\; -0.5x_1 + x_2 \le 1$

and $x_1, x_2 \ge 0.$

20. Max. $Z = 6x_1 - 2x_2$

s.t. $2x_1 - x_2 \leq 2,\ x_1 \leq 3$

and $x_1, x_2 \geq 0.$

21\. Max. $Z = 3x_1 + 4x_2$

s.t. $5x_1 + 4x_2 \leq 200,\ 3x_1 + 5x_2 \leq 150,\ 5x_1 + 4x_2 \geq 100,\ 8x_1 + 4x_2 \geq 80,\ x_1, x_2 \geq 0.$

22\. Max. $Z = x_1 + \frac{1}{2}x_2$

s.t. $3x_1 + 2x_2 \leq 12,\ 5x_1 \leq 10,\ x_1 + x_2 \geq 8,\ -x_1 + x_2 \geq 4,\ x_1, x_2 \geq 0.$

23\. Min. $Z = 2x_1 - 10x_2$

s.t. $x_1 - x_2 \geq 0,\ x_1 - 5x_2 \geq -5$ and $x, x_2 \geq 0.$

3

Feasible Solution of Transportation Problem

INTRODUCTION

A transportation problem is a special case of a linear programming problem. So the definition of B.F.S. is same as given earlier for L.P.P. But we find that in a transportation problem out of an unknowns there are only m + n – 1 basic variables. This happens due to redundancy in the constraints of the transportation problem. The condition $\sum a_i = \sum b_j$ can be used to reduce one constraint. This can be easily justified by proving the following theorem:

Theorem:

Out of (m + n) equations, there are only m + n – 1 independent equations in a transportation problem, m and n being the number of origins and destinations and any one equation can be dropped as the redundant equation.

Proof:

Consider m row equations and n – 1 column equations of the transportation problem as

$$\sum_{j=1}^{n} x_{ij} = a_i,\ i = 1, 2, \ldots, m \qquad \ldots(1)$$

and $$\sum_{i=1}^{m} x_{ij} = b_j,\ j = 1, 2, \ldots, n-1 \qquad \ldots(2)$$

Now adding m origin constraints given in (1), we get

$$\sum_{i=1}^{m}\sum_{j=1}^{n} x_{ij} = \sum_{i=1}^{m} a_i \qquad \ldots(3)$$

Also, adding (n – 1) destination constraints given in (2), we get

$$\sum_{j=1}^{n-1} \sum_{i=1}^{m} x_{ij} = \sum_{j=1}^{n-1} b_j \qquad ...(4)$$

Subtracting (4) from (3), we get

$$\sum_{i=1}^{m} \sum_{j=1}^{n} x_{ij} - \sum_{j=1}^{n-1} \sum_{i=1}^{m} x_{ij} = \sum_{i=1}^{m} a_i - \sum_{j=1}^{n-1} b_j$$

$$\text{or } \sum_{i=1}^{m}\left(\sum_{j=1}^{n} x_{ij} - \sum_{j=1}^{n-1} x_{ij} \right) = \sum_{j=1}^{n} b_j - \sum_{j=1}^{n-1} b_j \qquad [\because \sum a_i = \sum b_j]$$

or $\sum_{i=1}^{m} x_{in}$ = bn, which is the nth destination-constraint.

It follows that if m + n – 1 constraints are satisfied then the (m + n)th constraint will be automatically satisfied due to the condition $\sum a_i = \sum b_j$. Thus we have only (m + n – 1) linearly independent equations. Out of (m + n) equations, one (any) is redundant.

Hence the theorem is proved.

It indicates that B.F.S. will contain almost m + n – 1 positive variables, others being zero.

EXISTENCE OF AN OPTIMAL SOLUTION

Theorem:

There always exists an optimal solution to balanced transportation problem.

Proof:

We have $\sum_{i=1}^{m} a_i = \sum_{j=1}^{n} b_j$.

It follows that a feasible solution exists of the problem *i.e.*, $x_{ij} \geq 0$ for all i and j.

Form the constraints of the problem each $x_{ij} \leq \min(a_i, b_j)$.

Thus $0 \leq x_{ij} < \min(a_i, b_j)$ *i.e.*, the feasible region of the problem is non-empty, closed and bounded.

Hence, there exists an optimal solution.

SOLUTION OF A TRANSPORTATION PROBLEM

The solution of a transportation problem consists of the following two steps:

Step 1. To find an initial basic feasible solution.

Step 2. To obtain an optimal solution by making successive improvements to initial basic feasible solution (obtained in step 1) until no further decrease in the transportation cost is possible.

METHODS TO FIND AN INITIAL BASIC FEASIBLE SOLUTION.

Here we describe some simple methods to obtain the initial basic feasible solution.

Method 1: North-West Corner Rule

In this rule we have the following steps:

Step 1. Start with the cell (1, 1) at the north-west corner *i.e.*, the top-most left corner and allocate there maximum possible amount. Thus $x_{11} = \min(a_1, b_1)$.

Step 2.

(i) If $b_1 < a_1$ then $x_{11} = b_1$ and there is still some quantity available left in row 1. So move to the right hand cell (1, 2) and make the second allocation of amount $x_{12} = \min(a_1 - x_{11}, b_2)$ in the cell (1, 2).

(ii) If $b_1 > a_1$, then $x_{11} = a_1$ and there still some requirement left in column 1. So move vertically downwards to the cell (2, 1) and make the second allocation of amount $x_{21} = \min(a_1, b_1 - x_{11})$ in this cell.

(iii) If $b_1 = a_1$ then $x_{12} = 0$ or $x_{21} = 0$.

Start from the new north-west corner of the transportation table and allocate there as much as possible.

Step 3. Repeat steps 1 and 2 until all the available quantity is exhausted or all the requirement is satisfied.

The following example explains the method:

Example 1:

Find the initial basic feasible solution of the following transportation problem:

		To			
		W_1	W_2	W_3	Available
	F_1	2	7	4	5
	F_2	3	3	1	8
From	F3	5	4	7	7
	F_4	1	6	2	14
Requirement		7	9	18	34

Solution:

First we construct an empty 4 by 3 matrix complete with row and column requirements.

Start with the cell (1, 1) at the north-west corner (top-most left corner) and allocate it maximum possible amount. Thus $x_{11} = 5$ as minimum of $a_1 = 5$ and $b_1 = 7$ is 5.

		To			
		W_1	W_2	W_3	Available
	F_1	5 (2)			5
	F_2	2 (3)	6 (3)		8
From	F_3		3 (4)	4 (7)	7
	F_4			14 (2)	14
Requirement		7	9	18	

There is no amount left available at source 1 so in place of moving to right we move vertically downwards to the cell (2, 1) and allocate as much as possible there. The column 1 still needs the amount 2 and the amount 8 is available in row 2 so we allocate the maximum amount 2 to the cell (2, 1) *i.e.*, $x_{21} = 2$. Thus allocations for column 1 are complete. Now we move to the right of the cell (2, 1). Since the amount 6 is still available in row 2 and amount 9 is needed in column 2, so we allocate the maximum amount 6 is still available in row 2 and amount 9 is needed in column 2, so we allocate the maximum amount 6 in the cell (2, 2) *i.e.*, $x_{22} = 6$. This completes the allocations for row 2. Now we move vertically downwards to the cell (3, 2). In column 2 the amount 3 is still needed and in row 3 amount available is 7 so we allocate the maximum amount 3 in the cell (3, 2) *i.e.*, $x_{32} = 3$. Thus allocations for column 2 are complete.

Now the amount 4 is still available in row 3 and amount 18 is needed in column 3, so we move to the cell (3, 3) and allocate the maximum amount 4 to this cell *i.e.*, $x_{33} = 4$. Thus, there is no amount left available at source 3. Now we move downwards to the cell (4, 3). The amount 14 is still needed in column 3 and an equal amount 14 is available in row 4, so we allocate the amount 14 to cell (4, 3) *i.e.*, $x_{43} = 14$. The resulting feasible solution is shown in the above table. Allocations in the cells are in such a way that the total the total in each row and each column is the same as shown against the respective rows and columns

Multiplying each individual allocation by its corresponding unit cost in ( ), and adding, the total cost corresponding to this feasible solution is = 5.2 + 2.3 + 6.3 + 3.4 + 4.7 + 14.2 = Rs. 102.

Note:

In this method we always move to the right or down, so no loop can be formulated by drawing horizontal and vertical lines to the allocations. Also at each step (allocation) at least one row or column is discarded from further consideration, while the last allocation discards both a row and a column simultaneously, so we cannot get more than (m + n – 1) individual positive allocations. Thus *we always get a non-degenerate basic feasible solution by the north west corner rule.*

Method 2: Lowest Cost Entry (Matrix Minima) Method

In this method we have the following steps:

Step 1. Examine the cost matrix carefully and find the lowest cost. Let it be c_{ij}. Then allocate x_{ij} as much as possible in the cell (i, j). $x_{ij} = \min(a_1, b_j)$.

Step 2.

(i) If $x_{ij} = a_1$, then the capacity of the ith origin is completely exhausted. In this case cross out the ith row of the transportation table and decrease the requirement b_j by a_1. Now go to step 3.

(ii) If $x_{ij} = b_j$, then the requirement of jth destination is completely satisfied. In this case cross out the jth column of the transportation table and decrease a_i by b_j. Now go to step 3.

(iii) If $x_{ij} = a_i = b_j$, then either cross-out the ith row or jth column but not both. Now go to step 3.

Step 3. Repeat steps 1 and 2 for the reduced transportation table until all the available is exhausted or all the requirement is satisfied.

Note:

If the cell of lowest cost is not unique. We can select any one of these cells.

The method is well explained by taking the same numerical example as in method 1.

To

		W_1	W_2	W_3	
	F_1	(2)	2(7)	3(4)	5
From	F_2	(3)	(3)	8(1)	8
	F_3	(5)	7(4)	(7)	7
	F_4	7(1)	(6)	7(2)	14
		7	9	18	

First we write the cost and requirement matrix. We examine the cost matrix and find that there is lowest cost 1 in cell (2, 3) and in (4, 1). We choose any one of these. Say the cell (2, 3) and allocate the maximum possible amount 8 to this cell. This exhausts the availability from F_2. Now leaving the second row in the reduced transportation table we find that there is lowest cost 1 in cell (4, 1). Here we allocate the maximum possible amount 7. This satisfies the requirement of W_1. Leaving the first column, in the reduced transportation table we find the lowest cost 2 in the cell (4, 3). We allocate the maximum possible amount 7 to this cell. This exhausts the availability from F_4. Leaving the 4th row, in the reduced transportation table, we find that there is lowest cost 4 in cell (1, 3) and in cell (3, 2). We allocate the amount 3 in the cell (1, 3). This satisfies the requirement of W_3. In order to satisfy the availability of F_1 we allocate the amount 2 to the cell (1, 2). To complete the requirement of 9 units in column 2, we allocate the amount 7 to the cell (3, 2).

Thus we get the require B.F.S. shown in the above table.

The transportation cost = 2.7 + 3.4 + 8.1 + 7.4 + 7.1 + 72 = Rs. 83.

Method 3: Unit Cost Penalty Method (Vogel's Approximation Method).

In this method we have the following steps:

Step 1. Identify the smallest and next to smallest costs for each row of the transportation table. Find the difference between them for each row. Write these differences alongside the transportation table against the respective rows by enclosing them in parentheses. Write the similar differences for each column below the corresponding column. These are called 'penalties'.

Step 2. Now select the row or column for which the penalty is the largest. If a tie occurs, use any arbitrary tie breaking choice. Allocate the maximum possible amount to the cell with lowest cost in that particular row or column. Let the largest penalty correspond to ith row and let c_{ij} be the smallest cost in the ith row. Allocate the amount $x_{ij} = \min (a_i, b_j)$ in the cell (i, j).

Then we cross out the ith row or the jth column in the usual manner and construct the reduced matrix.

Step 3. Now compute the row and column penalties for the reduced transportation table and repeat the step 2. We continue this process until all the available quantity is exhausted or all the requirement is satisfied.

The method is well explained by taking the same numerical example as in method 1.

First we write the cost and requirement matrix and compute the penalties as follows:

	W_1	W_2	W_3	Available	Penalties
F_1	5(2)	(7)	(4)	5	(2)
F_2	(3)	(3)	(1)	8	(2)
F_3	(5)	(4)	(7)	7	(1)
F_4	(1)	(6)	(2)	14	(1)
Requirement	7	9	18		
Penalties	(1)	(1)	(1)		

We find that the maximum penalty (2) is associated with row 1 and row 2, so we may select any one of these. If we select row 1, then we allocate the maximum possible amount to the lowest cost cell in this row *i.e.*, cell (1, 1). Thus, x_{11} = min (5, 7) = 5. This exhausts the availability from F_1. So we cross the row 1. Leaving this row, the reduced cost and requirement matrix is as follows:

	W_1	W_2	W_3	Available	Penalties
F_2	(3)	(3)	8(1)	8	(2)
F_3	(5)	(4)	(7)	7	(1)
F_4	(1)	(6)	(2)	14	(1)
Requirement	2	9	18		
Penalties	(2)	(1)	(1)		

We note that the amount still needed to column 1 is 2.

Since the maximum penalty (2) is associated with row 1 and column 1 so we may select any one of these.

We select row 1 of this table and allocate the maximum possible amount 8 to the cell with cost 1 (lowest) in this row. This exhausts the availability from F_2 and leaves the requirement 10 for W_3. Again leaving row of F_2, the reduced transportation table is as follows:

	W_1	W_2	W_3	Available	Penalties
F_3	(5)	(4)	(7)	7	(1)
F_4	(1)	(6)	10(2)	14	(1)
Requirement	2	9	10		
Penalties	(4)	(2)	(5)		

In this table, the maximum penalty (5) is associated with column 3 so the maximum possible amount 10 is allocated to the cell with lowest cost 2 in this column. This completes the requirement of W_3. After leaving the column corresponding to W_3 the remaining table is as follows:

	W_1	W_2	Available	Penalties
F_3	(5)	7(4)	7	(1)
F_4	2(1)	2(6)	4	(5)
Requirement	2	9		
Penalties	(4)	(2)		

In this table, the maximum penalty (5) is associated with row 2, so the maximum possible amount 2 is allocated to the cell with lowest cost 1 in this row.

The remaining amount 2 available to F_4 is allocated to the cell with cost 6. In the last to meet the requirement of W_2 the amount 7 is allocated to the cell with cost 4.

Thus we get the required B. F. S. as shown in the table.

	W_1	W_2	W_3	Available
F_1	5(2)	(7)	(4)	5
F_2	(3)	(3)	8(1)	8
F_3	(5)	7(4)	(7)	7
F_4	2(1)	2(6)	10(2)	14
Requirement	7	9	18	

The total transportation cost

$$= 5.2 + 8.1 + 7.4 + 2.1 + 2.6 + 10.\ 2 = \text{Rs. } 80.$$

Note : If in the selected row or column the minimum cost is cost not unique, then allocate in that cell in which more allocations can be made.

Although, Vogel's method takes more time as compared to other methods, but it reduces the time in reaching the optimal solution. To obtain the optimal solution the students are advised to find the initial B. F. S. by Vogel's method.

TRANSPORTATION PROBLEM

The transportation problem can be described as follows.

Suppose that the factories F_i {i = 1, 2,..., m} called the origins or sources produce the non-negative quantities a_i = {i = 1, 2, – m} of product and non-negative quantities b_j (j = 1, 2,..., n) of the same product are required at other n places, called the destinations, such that the total quantity produced is equal to the total quantity required

i.e.,
$$\sum_{i=1}^{m} a_i = \sum_{j=1}^{n} bj \qquad ...(1)$$

Also support that C_{ij} is the cost of transportation of on unit from the ith source to the jth destination. Then the problem is to determine x_{ij}, the quantity transported from the ith source to the jth destination,

MATHEMATICAL FORMULATION OF TRANSPORTATION PROBLEM

Let there be m origins and n destinations (n may or may not be equal to m) with ith origin possessing ai units of a certain product and jth destination requiring b_j units of the same product.

Assume that the total available is equal to the total required

i.e., $$\sum_{i=1}^{m} a_i = \sum_{j=1}^{n} b_j \qquad ...(1)$$

Let c_{ij} be the cost of transportation of one unit product from ith origin to jth destination and x_{ij} be the quantity transported from the ith origin to the jth destination.

Then the problem is to determine non-negative (≥ 0) values of xij, satisfying the availability restrictions as well as the requirement restrictions, in such a way that the total transportation cost is minimized.

i.e., find x_{ij} (≥ 0) for i = 1, 2, ... , m; j = 1, 2, ... , n which minimize

$$Z = \sum_{i=1}^{m}\sum_{j=1}^{n} c_{ij}x_{ij} \qquad ...(2)$$

such that $$\sum_{j=1}^{n} x_{ij} = a_i,\ i = 1, 2, \ldots ,m \qquad ...(3)$$

and $$\sum_{i=1}^{m} x_{ij} = b_j,\ j = 1, 2, \ldots ,n \qquad ...(4)$$

The equations (3) and (4) may be called the row and column equations respectively.

In such a way that the total transportation cost $\sum_{i=1}^{m}\sum_{J=1}^{n} c_{iJ}x_{iJ}$ is minimized.

The transportation problem as described above can be represented in a tabular form as follows:

Destination→ Sources↓	w_1	w_2	w_j	w_n	Capacities of the sources
F_1	c_{11}	c_{12}	c_{1J}	c_{1n}	a_1
F_2	c_{21}	c_{22}	c_{2J}	c_{2n}	a_2
F_i	c_{i1}	c_{i2}	c_{ij}	c_{in}	a_i
F_m	c_{m1}	c_{m2}	c_{mJ}	c_{mn}	a_m
Requirements→	b_1	b_2	b_J	b_n	$\sum_{i=1}^{m} a_i \sum_{J=1}^{n} b_J$.

The calculations are directly on the transportation array given below which gives the curred trial solution.

Destination→ Sources↓	w_1	w_2	w_J	w_n	Capacities of the sources
F_1	x_{11}	x_{12}	x_{1j}	w_{1n}	a_1
F_2	x_{21}	x_{22}	x_{2j}	w_{2n}	a_2
F_i	x_{ij}	x_{i2}	x_{iJ}	x_{in}	a_i
F_m	x_{m1}	x_{m2}	x_{mJ}	x_{mn}	a_m
Requirements→	b_1	b_2	b_J	b_n	$\sum_{i=1}^{m} a_i \sum_{J=1}^{n} b_J$.

The above two tables can be combned together by writing the asts CiJ within the brocket ( ), as follow:

Destination→ Sources↓	w_1	w_2	w_J	w_n	the sources Capacities of
F_1	$x_{11}(c_{11})$	$x_{12}(c_{12})$	$x_{1J}(c_{1J})$	$w_{1n}(c_{1n})$	a_1
F_2	$x_{21}(c_{21})$	$x_{22}(c_{22})$	$x_{2J}(c_{2J})$	$w_{2n}(c_{2n})$	a_2
F_i	$x_{ij}(c_{12})$	$x_{i2}(c_{12})$	$x_{iJ}(c_{iJ})$	$x_{in}(c_{in})$	a_i
F_m	$x_{m1}(c_{m1})$	$x_{m2}(c_{m2})$	$x_{mJ}(c_{mJ})$	$x_{mn}(c_{mn})$	a_m
Requirements→	b_1	b_2	b_J	b_n	
$\sum_{i=1}^{m} a_i \sum_{J=1}^{n} b_J$.					

RELATION BETWEEN AN ASSIGNMENT PROBLEM AND TRANSPORTATION PROBLEM

An assignment problcm is a special case of the transportation problem when each origin is associated with one and only one destination. The numerical evaluations of such association are called 'effectiveness' in place of 'transportation costs'. In this case m = n, all a_i and b_j are unity and each x_{ij} is limited to one of the two values 0 and 1. In these circumstances exactly of x_{ij} can be non-zero (*i.e.*, 1), one for each origin and one for each dcstination.

FEASIBLE SOLUTION, BASIC FEASIBLE SOLUTION AND OPTIMUM SOLUTION OF TRANSPORTATION PROBLEM

Now we shall define few terms that are used in the transportation problem.

(i) **Feasible Solution (F.S.) :** A feasible solution to a transportation problem is a sets 0 non-negative individual allocations ($x_{ij} \geq 0$) which satisfies the row and column sum restriction [*i.e.*, equations (3) and (4) of 2].

(ii) **Basic Feasible Solution (B.F.S.) :** A feasible solution of an m by n transportation problem is said to be basic if the total number of positive allocations is equal to m + n – 1 *i.e.*, one less than the sum of the number of rows and columns.

(iii) **Optimum solution.** A feasible solution (not necessarily basic) is said to be optimum if it minimizes the total transportation cost.

EXISTENCE OF REUSABLE SOLUTION

Theorem:

A necessary and sufficient condition for the existence of feasible solution of a transportation problem is

$$\sum a_i = \sum b_j \quad (i = 1, 2, \ldots, m; j = 1, 2, \ldots, n).$$

Proof:

The condition is necessary: Let there exist a feasible solution to the transportation problem. Then

$$\sum_{j=1}^{n} x_{ij} = a_i, \ i = 1, 2, \ldots, m$$

and $\sum_{i=1}^{m} x_{ij}$ $j = b_j = 1, 2, \ldots, n.$

Summing over all i and j respectively, we get

$$\sum_{i=1}^{m}\sum_{j=1}^{n} x_{ij} = \sum_{i=1}^{m} a_i$$

and $$\sum_{j=1}^{n}\sum_{i=1}^{m} x_{ij} = \sum_{j=1}^{n} b_j$$

$$\Rightarrow \quad \sum_{i=1}^{m} a_i = \sum_{j=1}^{n} b_j.$$

The condition is sufficient: Let $\sum_{i=1}^{m} a_i = \sum_{j=1}^{n} b_j = k$ (say).

If $x_{ij} = \lambda_i b_j$ for all i and j, where $\lambda_i \neq 0$ is any real number, then

$$\sum_{j=1}^{n} x_{ij} = \sum_{j=1}^{n} \lambda_i b_j = \lambda_i \sum_{j=1}^{n} b_j = k\lambda_i$$

$$\Rightarrow \quad \lambda_i = \frac{1}{k}\sum_{j=1}^{n} x_{ij} = \frac{a_i}{k}.$$

Thus, $x_{ij} = \lambda_i b_j = \dfrac{a_i b_j}{k}$, for all i and j.

As $a_i \geq 0$, $b_j \geq 0$ so $x_{ij} \geq 0$ for all i and j.

Hence a feasible solution exists.

Note: Such a transportation problem in which $\sum a_i = \sum b_j$ is termed as balanced transportation problem. Hence a balanced transportation problem always has a F.S.

Difference between a Transportation and on Assignment Problem

An assignment problem is a special case of the transportation problem is which m = n, all the a_i and b_j are unity, and each x_{ij} is limited to one of the two value 0 and 1.

In these circumstances, exactly n of x_{ij} can be non-zero, one in each row of the array and one in each column showing that only one source (person) can be assigned to each destination (job).

Some important definitions.

1. *A feasible solution:* A feasible solution to a transportation problem is a set of non-negative individual all cations ($x_{ij} \geq 0$) which satisfies the row and column sum restrictions.
2. *Basic feasible solution:* A feasible solution of a m by n transportation problem is said to be a basic feasible solution if the total number of positive allocations x_{ij} is exactly equal to m + n – 1, *i.e.* one less than the sum of the number of rows and columns.
3. *Optimal solution:* A feasible solution (necessarily basic) is said to be optimal it minimizes the total transportation cost.
4. *Now-degenerate basic feasible solution:* A feasible solution of m by n transportation problem is said to be non degenerate basic feasible solution if.

   (i) Total number of positive allocations is exactly equal to (m + n – 1).

   (ii) these allocations are in independent portions.

DEGENERACY IN TRANSPORTATION PROBLEM

We recall that a B.F.S. to an m-origin and n-destination transportation problem will contain almost (m + n – 1) independent non-zero allocations. If this number is exactly (m + n – 1), the B.F.S. is said to be non-degenerate otherwise it is said to be a degenerate one. Thus degeneracy in transportation problem occurs whenever the number of independent individual allocations is cell than (m + n – 1).

Degeneracy in transportation problem can occur in two ways:

(i) B.F.S. may be degenerate from the initial stage onward.

(ii) It may become degenerate at any intermediate stage, when the selection of one entering cell empties two or more pre-occupied cells simultaneously.

In such cases, to resolve degeneracy, we allocate an extremely small amount (close to zero) to one or more empty cells of the matrix (generally lowest cost cells if possible), so that the total number of occupied (allocated cells become (m + n – 1) at independent positions.

The extremely small quantity usually denoted by Δ (delta) or $\in$ (epsilon) satisfies the following conditions:

1. $\Delta < x_{ij}$ for $x_{ij} > 0$
2. $x_{ij} + \Delta = x_{ij} = x_{ij} - \Delta, x_{ij} > 0$
3. $\Delta + 0 = \Delta$.
4. If there are more than one Δ's introduced in the solution, then.
   (i) if Δ, Δ' are in the same row, Δ < Δ' when Δ is to the left to Δ' and
   (ii) if Δ, Δ' are in the same column, Δ < Δ' when Δ is above Δ'.

Above rules show that even after introducing Δ, the original solution of the problem is not changed. It is merely a technique to apply the optimality test. As Δ has no physical significance, ultimately it is to be omitted.

UNBALANCED TRANSPORTATION PROBLEM

If in a transportation problem, the sum of all available quantities is not equal to the sum all requirements i.e., if $\sum_{i=1}^{m} a_i \neq \sum_{j=1}^{n} b_j$ *then such problem is called an unbalanced transportation problem.*

An unbalanced transportation problem may occur in two different forms:

(i) *Shortage in availability, i.e.,* $\sum a_i < \sum b_j$: To modify this type of unbalanced transportation problem to balanced type we introduce a dummy source row in the transportation table. The unit transportation costs from this dummy source to any destination are all set equal to zero. The availability at this dummy source is assumed to be equal to the difference $\sum b_j - \sum a_i$.

(ii) *Excess of availability, i.e.,* $\sum a_i > \sum b_j$. To modify this type of unbalanced transportation problem to balanced type we introduce a dummy destination column in the transportation table. The unit transportation costs from this dummy destination from any source are all set equal to zero. The requirement at this dummy destination is assumed to be equal to the difference $\sum a_i - \sum b_j$.

SOLUTION OF A TRANSPORTATION PROBLEM

The solution (optimal) of a transportation problem consists of the following two steps.

Step 1. To find on initial basic feasible solution.

Step 2. To obtain an optimal solution making improvements to initial basic passible solution until no further decrease in the transportation cot is possible.

To find an initial feasible solution.

There are several methods for finding the initial feasible solution of the given transportations problem. Here we describe the following three methods

Method 1: North-West Corner Rule

By this rule we allocate a set of allocations in the cells so that the row totals and columns totals will be as indicted before each, as follows.

(i) Start with the cell (1, 1) at the north-west corner *i.e.* the top most left corner and allocate it maximum possible amount.

(ii) Then move to the right hand cell (1, 2) if there is sill any available quantity left, dowries move to the down cell (2, 1) and allocate it maximum possible amount.

(iii) Repeat the step (ii) again and continue until all the available quantity is exhausted.

The method is well explained by toping the following numerical example:

To

		W_1	W_2	W_3	Supply
From	F_1	(2)	(7)	(9)	5
	F_2	(3)	(3)	(1)	8
	F_3	(5)	(4)	(7)	7
	F_4	(1)	(6)	(2)	14
Demand		7	9	18	34

To

		W_1	W_2		
From	F_1	5 (2)			$5 = a_1$
	F_2	2 (3)	6 (3)		$8 = a_2$
	F_3		3 (4)	4 (7)	$7 - a_3$
	F_4			14 (2)	$14 = a_4$
		7 b_1	$9 = b_2$	$18 = b_3$	

(i) We start with the topmost left corner and allocate is maximum possible amount 5. (Since mini of a_1 = 5 and b_1 = 7).

(ii) Since there is no amount left available at source 1 so we move downwards to the cell (2, 1), in place of moving to right, and allocate it maximum possible amount. Since the column 1 still need the amount 2 and the amount 8 is available in row 2 so we allocate the maximum amount 2 to this cell (2, 1).

Now the allocation for column 1 is complete, so we move to the right of this cell. Since the amount 6 is still available in row 2 and amount 9 is needed in column 2, so we allocate the maximum amount 6 in this cell (2, Thus, allocation for row 2 is complete, so we move downwards to the cell (3, 2). Since the amount 3 is still needed in column 2 and amount 7 is available in row 3, so we allocate the maximum amount 3 in this cell (3, 2). Thus allocation for column is complete, so we move to the right of this cell. Since the amount 4 is still available in row 3 and amount 19 is needed in column 3, so we allocate the maximum amount 4 to the cell (3, 3).

Thus the requirements of the row 3 is complete, so we more to the downgrade cell (4, 3). Since amount 14 is still available in row 3 and an equal amount 14 is needed in row 4, so we allocate this amount 14 to this cell (4, 3). This is complete the allocation.

In the end it may be checked that the sum of rows and columns are as needed. On multiplying each individual allocation by its corresponding unit in ( ), and adding, the total transportation cost do this F. S is

$= 5 \times 2 + 2 \times 3 + 6 \times 3 + 3 \times 4 + 4 \times 7 + 14 \times 2$

= Rs. 102.

In this north-west corner rule we more to the right or down, so no loop can be formulated here by drawing horizontal and vertical lines to the allocation. Also at each step at least one row or column is discarted from further consideration and at the last allocation both rows and columns are discarded.

So we cannot get more than (m + n – 1) individual positive allocation by this rule. Thus we always get a non degenerate basic feasible solution by this north-west corner rule.

Method 2: Lowest Cost Entry Method (Layman or Method of Matrix Minima)

In this we write the costs and the requirement matrix. The cost are written within brackets ( ). Now we examine the cost matrix carefully and choose the cell with lowest cost and allocate there as much as passible. If such cell of lowest cost in not unique, we can select any one of these cells. Again we examine the cost matrix and select the cell with the lowest cost (the cell in which allocation has been made is not considered) and allocate there as much as possible we continue this process unit all the available quantify is exhausted.

The method is well explained by taping the same number example in method 1.

First we write the cost and requirements matrix as following.

		To			
		W_1	W_2	W_3	
	F_1	(2)	2(7)	3(4)	5
From	F_2	(3)	(3)	8(1)	8
	F_3	(5)	7(4)	(7)	7
	F_4	7(1)	(6)	7(2)	14
		7	9	18	

Examine the cost matrix in above table we find there is lowest cost 1 in cell (2, 3) and in (4, 1) we choose any one of these say the cell (2, 3) and allocate the maximum amount 8 to this cell Leaving this cell we find that there is lowest best 1 in cell (4, 1) where we allocate the maximum amount 7. Continuing in this way we get the required feasible solution shows in always table.

The total transportation cost to this F. S. is

$$= 2 \times 7 + 3 \times 4 + 8 + 1 + 7 \times 4 + 7 \times 1 + 7 \times 2$$

$$= \text{Rs. } 83.$$

This cost is less than the cost associated with the F.S. obtained by north west corner rule.

The initial feasible solution obtained by this method usually gives a lower transportation cost than that obtained is north west corner rule.

Method 3: Unit Cost Penalty Method (or Vogel's Approximation Method)

In this method we write the differences of the smallest and the second smallest costs in each column, below the correspond column and write the similar differences of each row to the right of the corresponding row. These individual different can be thought of penalty for making allocations in the second lowest cost cell instead of lowest cost cell in each row or column.

Now we select the row or column for which the penalty is the largest and allocate the maximum possible amount to the cell with lowest cost in that particular row or column. If there are more than one largest penalty rows or column, then we may select any one of them. Then we cross that row or column in which the requirement has been satisfied and construct the reduced matrix. We continue this process on the reduced matrices till all the allocations have been made.

The method is well explain by taking the numerical examples as in method I.

First we write the cost and requirement matrix as follows:

	W_1	W_2	W_3	Available	penalties
F_1	5(2)	(7)	(4)	5	(2)
F_2	(3)	(3)	(1)	8	(2)
F_3	(5)	(4)	(7)	7	(1)
F_4	(1)	(6)	(2)	14	(1)
Demand	7	9	18		
Penalties	(1)	(1)	(1)		

Since the maximum penalty (2) is associated with row 1 and row 2, so we may select any one of these. Suppose we select row 1 then we allocate the maximum possible amount 5 to the cell (1, 1) with lowest cost and cross this row 1.

The first reduced matrix after leaving row with remaining demand and available is and follows.

(Note that the amount 5 has been allocated to the column 1, so the amount still needed to column 1 is 2)

	W_1	W_2	W_3	Available	Penalties
F_2	(3)	(3)	8(1)	8	(2)
F_3	(5)	(4)	(7)	7	(1)
F_4	(1)	(6)	(2)	14	(1)
Demand	2	9	18		
Penalties	(2)	(1)	(1)		

Since the maximum penalty (2) is associated with row 1 above table, so the maximum possible amount 8 is allocated to the cell (1, 3) with lowest cot in this row.

The second reduced matrix after leaving row 1 of the matrix in above table with remaining demands and available is as follows

	W_1	W_2	W_3	Available	Penalties
F_3	(5)	(4)	(7)	7	(1)
F_4	(1)	(6)	10(2)	14	(1)
Demand	2	9	10		
Penalties	(4)	(2)	(5).		

Since the maximum penalty (5) is associated with column 5 of matrix in above table, so the maximum possible amount 10 is allocated to the cell (2, 3) with lowest cost in this column.

The third reduced matrix after learning column 3 of the matrix in above table with remaining demands and availables is as follows:

	W_1	W_2	Available	Penalties
F_3	(5)	7(4)	7	(1)
F_4	2(1)	2(6)	4	(5)
Demand	2	9		
Penalties	(9)	(2)		

Since the maximum penalty (5) is associated with row 2, so the maximum possile amount 2 is allocated to the cell (2, 1) with lowest cost in this row. The remaning amount 2 in row second is allocated to the cell (2, 2) and then we allocate the renaming amount 7 to the cell (1,

Thus we get the required feasible solution as shown in below table:

	W_1	W_2	W_3	Available
F_1	5(2)	(1)	(4)	5
F_2	(3)	(3)	8(1)	8
F_3	(5)	7(4)	(7)	7
F_4	2(1)	2(6)	10(2)	14
Demand	7	9	18	

The total transportation cost to this F.S.

$= 5 \times 2 + 8 \times 1 + 7 \times 4 + 2 \times 1 + 2 \times 6 + 10 \times 2$

= Rs. 80.

Which is less than the cost associated with the feasible solution obtained by the previous method 1 and 2.

Theorem:

It we have a B.F.S consisting of $(m + n - 1)$ independent positive allocations, and a set of arbitrary numbers u_i and v_j, $i = 1, 2,..., m$,

$j = 1, 2, n$, such that

$$c_{rs} = u_r + v_s$$

For all occupied cells (r, s), then the evaluation d_{ij} corresponding to each empty cell (i, j) is given by

$$d_{ij} = c_{ij} - (u_i + v_j).$$

Proof:

Mathematically a transportation problem is to determine $x_{ij} \geq 0$, which minimize.

$$Z = \sum_{i=1}^{m} \sum_{J=1}^{n} c_{iJ} . x_{iJ} \quad ...(1)$$

subject to the restrictions

$$\sum_{J=1}^{n} x_{ij} = a_1 \text{ or } 0 = a_i - \sum_{J=1}^{n} x_{ij} \quad ...(2)$$

for $i = 1, 2, ... ,m$

$$\sum_{i=1}^{m} x_{iJ} = b_j \text{ or } 0 = b_j - \sum_{i=1}^{m} x_{iJ} \quad ...(3)$$

for j = 1, 2, ..., n.

Multiplying (2) by u_i (i = 1, 2..., m), 3 by v_j (= 1, 2...,n) and adding to (11) we have

$$Z = \sum_{i=1}^{m}\sum_{J=1}^{n} c_{iJ}x_{iJ} + \sum_{i=1}^{m} u_i\left(a_i - \sum_{J=1}^{n} x_{iJ}\right) + \sum_{J=1}^{n} v_J\left(b_j - \sum_{i=1}^{m} x_{iJ}\right)$$

$$\text{or } Z = \sum_{i=1}^{m}\sum_{j=1}^{n}[a_{ij} - (v_i + v_j)]x_{ij} + \sum_{i=1}^{n} v_i a_i + \sum_{j=1}^{n} v_j b_j \qquad \text{...(4)}$$

But it is given that for all occupied cells (cells with positive allocations)

$$c_{rs} = u_r + v_s \qquad \text{...(5)}$$

∴ In the objective function (4), all the terms of positive allocation vanish as this coefficients are zero.

For thus F.S the value of objective function (4), reduce to

$$Z = \sum_{i=1}^{m} u_i a_i + \sum_{j=1}^{n} v_j b_j \qquad \text{...(6)}$$

To prove the required result, let us first determine the cell evaluation for the empty cells (i_1 j) when we allocate + 1 unit to this empty cell the number of positive allocations become (m + n), hence they become independent in position and so a closed loop can be formed. Let us closed loop formed by Joining this cell (i, j) to the occupied cells be as shown in fig (1).

Here cell (i, s), r, s) and (r, j) ore occupied cells

∴ $c_{is} = u_i + v_s$, $c_{rs} = u_r + v_s$, $c_{rj} = u_r + v_j$ when we allocate + 1 unit to the empty cell (i, j), then to maintain the row and column sums unchanged we shall decrease the individual allocation at the cell (i, s) by 1, increase that at the less (r, s) by 1 and decrease (i, j) that at cell (r, T) by 1. That the c_{ij} allocation at the occupied cells (i, s), (r, s), (r, T) are changed. Thus, the cell evaluation d_{ij} corresponding to this empty cell (i, j) is given by d_{ij} = Difference in the total cost for the new solution and the original one

$$= c_{iT} - c_{is} + c_{rs} - c_{rj}$$

$$= c_{ij} - (u_i + v_s) + (u_r + v_s) \;\; (u_r + v_j)$$

$$\text{or } d_{ij} = c_{ij} - (u_i + v_j)$$

This gives the evaluations d_{ij} connecting empty cell (i, j) to the occupied cells by a square shaped loop. By similar reasoning, we way generalize it to the occupied cells.

Thus, we conclude that the evaluation d_{ij} corresponding to each empty cell (i, j) is given by

$$d_{ij} = c_{ij} - (u_i + v_T)$$

This completes the proof of the theorem.

Computational Procedure of Optimal Test

After getting the initial B.F.S of a transportation problem, we test this solution for optometry as follows.

1. For a B.F.S. we determine a set of (m + n) numbers

   $u_i \neq i = 1, 2, \dots, m$

   $v_j \neq J = 1, 2, \dots, n$

   Such that for each occupied cell (r, s)

   $c_{rs} = u_r + v_s$

   For this we assign an arbitrary value to one of the u_i's or v_j's then the rest (m + n – 1) of them can early resolved algebraically from the relation $c_{rs} = u_r + v_s$, for occupied cells. Generally we chooses that u_i or $v_j = 0$ for which the corresponding row or column have the maximum number of undivided allocations.

2. Then we calculate the cell evaluations d_{ij} for each unoccupied cell (i, j) by wring the formula $d_{ij} = c_{ij} - (u_i + v_j)$.

3. Then we examine the matrix of cell evaluations and crackled that

   (i) If all $d_{ij} > 0$, the solution under test is optimal and unique.

   (ii) If all $d_{ij} \geq 0$, with at cost one $d_{ij} = 0$, then the solution under test is optimal and alternative optimal solution exists.

   (iii) If at test one $d_{ij} < 0$, then the solution is not optimal. In the last case we proceed to the next step 4.

4. If at least one $d_{ij} < 0$, in step 3, then we form a new B.F.S. In the next B.F.S we give maximum allocation to the cell for which d_{ij} is minimum and negative, by making an occupied cell empty.

5. Then we repeat the steps 1 to 3 to test the optimality to this new B.F.S.

EXERCISES

1. Explain briefly the step-wise description of the computational procedure for solving the transportation problem.

2. Obtain an initial basic feasible solution to the following transportation problem. Is this solution an optimal solution? If not, obtain the optimal solution.

	W_1	W_2	W_3	W_4	a_i
F_1	19	30	50	10	7
F_2	70	30	40	60	9
F_3	40	8	70	20	18
$b_j \rightarrow$ 5	8	7	14		

3. Solve the transportation problem:

	D_1	D_2	D_3	D_4	Available
O_1	1	2	1	4	30
O_2	3	3	2	1	50
O_3	4	2	5	9	20
Required	20	40	30	10	100 Total

4. Explain how it is tested whether a B.F.S. of a transportation problem is optimal or not.

5. How does the problem of degeneracy arise in a transportation problem?

6. Explain how to solve the degeneracy in transportation problems.

7. Explain the difference between a transporting problem and an assignment problem.

8. If all the sources are emptied and all the destinations are filled show that $\sum a_i = \sum b_j$ is a necessary and sufficient condition for the existence of a feasible solution to a transportation problem.

9. Prove that there are only m + n – 1 independent equations in a transportation problem, m and n being the no. of origins and destinations respectively and that any one equation can be dropped as the redundant equation.

10. Explain the north-west corner rule for obtaining an initial basic feasible solution of a transportation problem.
11. Explain the lowest cost entry method for obtaining an initial basic feasible solution of a transportation problem.
12. Explain Vogel's Approximation Method of solving a transportation problem.
13. Is $x_{13} = 50, x_{14} = 20, x_{21} = 55, x_{31} = 30, x_{32} = 35, x_{34} = 25$ an optimum solution of the following transportation problem?

					Available units
	6	1	9	3	70
	11	5	2	8	55
	10	12	4	7	90
Required units	85	35	50	45	

If not, modify it to obtain an optimum basic feasible solution.

14. What is Transportation Problem? Give the mathematical formulation of transportation problem.

4

Convex Sets and Their Properties

INTRODUCTION

Definition (Convex Set): *A subset SCR^n is said to be convex is convex self all their properties for each pair of points x, y in S, the line segment [x : y] is contained in S.*

In symbols, a subset SCR^n is convex iff

$x, y \varepsilon S \Rightarrow [x : y] CS.$

We illustrate with the help of some examples.

An Open Set. *A set S is said to be an open set if it contains only the interior points.*

A Closed Set. *A set S is said to be a closed set if it contains all its boundary points.*

Lines, In E^n. *The line through the two points x_1 and x_2, $x_1 \neq x_2$ is defined to be the set of points*

$X = \{x{:}x = \lambda x_1 + (1 - \lambda) x_2, \text{ for all real } \lambda\}.$

Line Segments. *In E^n, the line segment joining two points $x1$ x_2 is defined to be the set of points*

$X = \{x{:}x = \lambda x_1 + (1 - \lambda) x_2,$

$0 \leq y \leq 1\}.$

Note that the restriction $0 \leq y \leq 1$ restricts the point x to lie within the in joining the points x_1 and x_2. Line segment of x_1, x_2 is also denoted by $[x_1 : x_2]$.

Note that the restriction $0 \le l \le 1$ restricts the point x to lie within the line joining the points x_1 and x_2. Line segment of x_1, x_2 is also denoted by $[x_1 : x_2]$.

Hyperplane. ***A hyperplane*** *is defined as the set of points satisfying*

$$c_1 x_1 + c_2 x_2 + \dots + c_n x_n = z \text{ (not all } c_i = 0)$$

or $$c x = z$$

for prescribed values of $c_1, c_2, \dots, c_n$ *and z.*

For optimum value of z this hyperplane is called optimal hyperplane.

The vector c is called a *vector normal to the hyperplane* and $\pm \dfrac{c}{|c|}$ are called unit normal.

It can be easily seen that hyperplanes are closed sets.

A hyperplane divides the whole space E^n into three mutually disjoint sets given by

$$X_1 = \{x : c x > z\}$$

$$X_2 = \{x : c x = z\}$$

$$X_3 = \{x : c x < z\}.$$

The sets X_1 and X_2 are called open half spaces.

The sets $\{x : c x \le z\}$ and $\{x : c x \ge z\}$ are called closed half spaces.

Note: The objective function of a L.P.P. represents a hyperplane and each constraint (sign $\le$, $\ge$) is a closed half space produced by the hyperplane given by the constraint by taking (=) sign in place of $\le$ and $\ge$.

Parallel Hyperplanes. *Tow hyperplanes* $c_1 x = z_1$ *and* $c_2 x = z_2$ *are said to be parallel if they have the same unit normals i.e., if* $c_1 = \lambda c_2$ *for some* λ, λ *being non-zero.*

CONVEX COMBINATION

A convex combination *of a finite number of points* $x_1, x_2, \dots, x_n$ *is defined as a point*

$$x = \lambda_1 x_1 + \lambda_2 x_2 + \dots + \lambda_n x_n,$$

where λ_i *is real and* ≥ 0, $\forall\, i$ *and* $\sum_{i=1}^{n} \lambda_i = 1$.

The convex linear combination of two points x_1 and x_2 is given by

$$x = \lambda_1 x_1 + \lambda_2 x_2, \text{ s.t. } \lambda_1, \lambda_2 \geq 0, \lambda_1 + \lambda_2 = 1.$$

It can also be written as

$$x = \lambda x_1 + (1 - \lambda) x_2, 0 \leq 1 \leq 1.$$

This shows that the line segment of the two points x_1 and x_2 is nothing but the set of all possible convex combinations of the two points x_1 and x_2.

EXTREME POINT OF A CONVEX SET

A point x in a convex set C is called on **extreme point** *if x cannot be expressed as a convex combination of any two distinct points x_1 and x_2 in C.*

In other words, a point x in a convex set C is an extreme point of C if it docs not lie on the line segment of any two points, different from x in the set.

Mathematically, a point x is an extreme point of a convex set if there do not exist other points x_1, x_2 ($x_1 \neq x_2$) in the set such that

$$x = \lambda x_1 + (1 - \lambda) x_2, 0 > \lambda < 1.$$

For example : The set $C = \{(x_1, x_2) : x_1^2 + x_2^2 \leq 1\}$ is convex. Every point on the circumference is an extreme point. Thus a convex set may also have infinite number of extreme points.

The polygons which are convex sets have the extreme points as their vertices.

Obviously, an extreme point is a boundary point of the set.

It is important to note that all boundary points of a convex set are not necessarily extreme points.

A point of C which is not an extreme point, is referred as an internal point of C.

CONVEX HULL

The **convex hull** *C(X) of any given set of points X is the set of all convex combinations of sets of points from X.*

In other words, the intersection of all convex sets, containing $X \subset E^n$, is called the convex hull of X and is denoted by $< X >$. Thus, the convex hull of a set $X \subset E^n$, is the smallest convex set containing X.

Example :

If X is the set of eight vertices of a cube, then the convex hull C(X) is the whole cube.

ANALYTICAL METHOD

The L.P. problems having more than two variables cannot be solved by Graphical method. In such cases, the analytic solution (trial and error method) may seem to be useful. For solving a L.P.P. Simplex method is the most powerful technique (discussed in the next chapter).

Extreme point theorem states that an optimal solution to a L.P.P. occurs at one of the vertices of the feasible region.

Since the vertices of the feasible region correspond to the basic feasible solutions., the objective function is optimal at least at one of the basic solutions. Some of the vertices may be infeasible which are dropped from consideration.

Consider the following L.P.P.

Maximize $Z = -0.10x_1 + 0.50x_2$

subject to the conditions $2x_1 + 5x_2 \leq 80$

$x_1 + x_2 \leq 20$

$x_1, x_2 \geq 0.$

Introducing the slack variables, the given L.P.P. reduces to the following form:

Maximize $Z = -0.10x_1 + 0.50x_2 + 0.x_3 + 0.x_4$

subject to $2x_1 + 5x_2 + x_3 = 80$

$x_1 + x_2 + x_4 = 20$ ...(1)

$x_1, x_2, x_3, x_4 \geq 0.$

We know that for a system of m equations in n unknowns (n > m), a solution is called a basic solution if in it at least (n – m) variables have the value zero. This gives a vertex.

Taking (n – m) variables equal to zero at a time and solving the resulting system of equations we get a basic solution.

Here n = 4 (the number of variables) and m = 2 (the number of equations). Out of 4 variables 2 can be taken in $^4C_2 = 6$ number of ways. Hence at the most there will be 6 basic solutions which can be obtained as follows:

1. Taking $x_1 = x_2 = 0$, system (1) gives $x_3 = 80, x_4 = 20$.
2. Taking $x_3 = x_4 = 0$, system (1) gives $x_1 = 20/3, x_2 = 40/3$.
3. Taking $x_2 = x_3 = 0$, system (1) gives $x_1 = 40, x_4 = -20$.
4. Taking $x_2 = x_4 = 0$, system (1) gives $x_1 = 20, x_3 = 40$.
5. Taking $x_1 = x_4 = 0$, system (1) gives $x_2 = 20, x_3 = -20$.
6. Taking $x_1 = x_3 = 0$, system (1) gives $x_2 = 16, x_4 = 4$.

Since basic solutions 3 and 5 so not satisfy the non-negative restriction of the variables so these do not give the B.F. solutions and so are dropped from the consideration.

Thus the B.F. solutions of the given L.P.P. are given by 1, 2, 4 and 6. For these B.F. solutions the values of the objective function Z are respectively

$$Z_1 = 0,\ Z_2 = 6,\ Z_4 = -2 \text{ and } Z_6 = 8.$$

Since maximum $Z = Z_6 = 8$, so the optimal solution to the given L.P.P. is given by $x_1 = 0$, $x_2 = 16$ and max. $Z = 8$.

The same solution is obtained earlier also by graphical method.

Here it is important to note that the B.F. solutions 1, 2, 4 and 6 exactly coincide with the corner points of the feasible region and one of these corner points gives the optimal solution.

Drawbacks of Analytical Method

(i) For large values of m and n, it is extremely difficult and time consuming to solve various sets simultaneous equations.

(ii) Some of the sets give infeasible also which are dropped from consideration.

There is no technique to detect all such sets and not to solve them at all.

(iii) We observe that the value of Z changes from 0 to 6 to – 2 to 8 *i.e.*, there are ups and downs.

VECTORS

Definition 1:

(n-vectors) An N-vector a is an ordered n-taple of real numbers, written in the form of a column,

$$a = \begin{bmatrix} a_1 \\ a_2 \\ \cdots \\ \cdots \\ \cdots \\ a_n \end{bmatrix} \text{ or } [a_1, a_2, \ldots, an]$$

The numbers a_i (i = 1, 2, ... n) are called its components.

The set of all n-vectors is denoted by the symbol R^n.

The notation $[a_1, a_2, \ldots a_n]$ for an n-vector indicates and n-typle written in the form of a column. It should not be confused with the notation $(a_1, a_2, \ldots, a_n)$ for a paint.

However, there is a connection, in that the n-vector a, that corresponds to the paint $A = (a_1, a_2, \ldots, a_n)$, is the vector.

$$a = [a_1, a_2, \ldots, a_n]$$

Definition 2:

(Equality of two n-vectors). Two n-vectors $a = [a_1, a_2, \ldots, a_n]$ and $b = [b_1, b_2, \ldots b_n]$ are said to equal if and only if $a_i = b_i$ for all i = 1, 2, ... ,n note that if a = b, then b = a.

Definition 3:

(Sum of two vectors). The sum of two n-vectors $a = [a_1, a_2, \ldots a_n]$ and $b = [b_1, b_2, \ldots, b_n]$, written as a + b, is the n-vector

$$a + b = [a_1 + b_1, a_2 + b_2, \ldots, a_n + b_n].$$

It is easy to verity the following properties of vector addition:

(i) for all a, b $\in R^n$,

$$a + b = b + a$$

(ii) for all a, b c $\in R^n$,

$$a + (b + c) = (a + b) + c$$

(iii) There exist the zero vector 0 = [0, 0, ... ,0] such that for all a $\in R^n$,

$a + 0 = 0 + a = a.$

Definition 4:

(Multiplication of a vector by a scalar). The Scalar Multiple of an n-vector, $a = [a_1, a_2, \ldots, a_n]$ by a scalar λ, written as λ a, is defined to be the n-vector

$$\lambda a = [\lambda a_1, \lambda a_2, \ldots, \lambda a_n]$$

Thus $\lambda a = -a$, if $\lambda = -1$

and $\lambda a = 0$ if $\lambda = 0$.

The operation of finding a scalar multiple satisfies the following properties.

(i) For all a, b $\in R^n$, and scalar c,

$c (a + b) = c_a + c_b$

(ii) For any vector a $\in R^n$, and scalar c_1, c_2

$(c_1 + c_2) a = c_{1a} + c_{2a}$

(iii) For any vector a $\in R^n$, and scalars c_1, c_2

$c_2 (c_1 a) = (c_2 c_1) a = c_1 (c_2 a)$

(iv) For any vector a $\in R^n$,

$1 a = a$

We omit the verification, which is straight forward.

Definition 5:

(Inner product or scalar preadult). Given two n-vectors, $a = [a_1, a_2, \ldots, a_n]$ and $b = [b_1, b_2, \ldots b_n]$ the INNER (SCALAR) PRODUCT of a and b, to be denoted by a. b is the scalar

$$a. b = a_1 b_1 + \ldots + a_n b_n$$

$$\text{or } a. a = \sum_{i=1}^{\lambda} a_i b_i$$

It is easy to verity the following:

(i) a. b = b. a

(ii) (a + b).c = a.c + b.c

(iii) a.l (b) = λ. (a. b) = (λa.). b

Definition 6:

(The length of a vector). The length (or the norm) or a vector a ∈ R^n to be denoted by | a | is defined as $|a| = \sqrt{a.a}$

$$|a| = \sqrt{a_1^2 + a_2^2 + ... a_n^2}.$$

Definition 7:

Unit vector. A vector n is acid to be unit sector if | n | = 1

Definition 8:

Angle between two vectors. The angle θ between two vectors is defined

$$\cos\theta = \frac{a.b}{|a|.|b|} \quad 0 \le \theta \le \Pi.$$

LINEAR COMBINATION OF VECTORS

Given two vectors a and b, we can always perform operations of addition and scalar multiplication to obtain vectors of the form a + b, a – b, 5a + 3b, 2a – 7b, etc.

Such vectors are called linear combinations of a and b. In general, for any scalars λ and μ, the vector λa + μb is a linear combination of a and b.

Thus, if $\{a_1, a_2, ... ,a_k\}$ is a set of k vectors from R^n and $\{\lambda_1, ... ,\lambda_2, \lambda_k\}$ is a set of k scalars, then the vector

$$\lambda_1 a_1 + \lambda_2 a_2 + ... + \lambda_k a_k$$

is known as a linear combination of the given set of vectors

$$a_1, a_2, ... , a_k.$$

Definition 1:

(Linear dependence and independence of vectors). *A set of vectors $a_1, a_2, ... ,a_k$ of R^n, is said to be LINEARLY DEPENDENT if there exist scalars $\lambda_1, \lambda_2, ... ,\lambda_k$ not all zero, such that*

$$\lambda_1 a_1 + \lambda_2 a_2 + ... + \lambda_k a_k = 0.$$

When it is not possible to have the above equality without all l_i's being zero, then the given set of vectors $a_1, a_2, ... ,a_k$ is said to be Linearly independent.

Illustration 1:

The vectors $a_1 = [1, 2]$ and $a_2 = [2, 4]$ are linearly dependence since there exist $\lambda_1 = 2$ and $\lambda_2 = -1$ for which

$$\lambda_1 a_1 + \lambda_2 a_2 = 0.$$

Illustration 2:

The vectors $c_1 = [1, 0, 0]$, $c_2 = [0, 1, 0]$ and $c_3 = [0, 0, 1]$ are linearly independent since

$$\lambda_1 c_1 + \lambda_2 c_2 + c_3 \lambda_3 = 0$$

$$\Rightarrow \lambda_1 [1, 0, 0] + \lambda_2 [0, 1, 0] + [0, 0, 1] = [0, 0, 0]$$

$$\Rightarrow [\lambda_1, \lambda_2, \lambda_3] = [0, 0, 0]$$

$$\Rightarrow \lambda_1 = 0, \lambda_2 = 0 \text{ and } \lambda_3 = 0,$$

Note: If a set of vectors is linearly independent, then any sudsier of it is also linearly independent. If a set of vectors is linearly dependent, then a superset of it is also linearly dependent.

Definition 2:

(Basis set). *A linearly independent set of vectors $a_1, a_2, \dots a_n$, of R^n, having the property that any vector of R^n can be expressed as a linear combination of these vectors, is called Basis set for R^n.*

Theorem 1:

The set of n vector e_i, $i = 1, 2, \dots, n$ where e_i has i th component equal to 1, and all other components zeroes, forms a basis for R^n.

Proof:

We shall show that the vectors $e_1, e_2, \dots, e_n$ of R^n are linearly independent and that any vector of R^n can be expressed as linear combination of these unit vectors.

Consider

$$\sum_{i=1}^{n} \lambda_i e_i = 0, \text{ where } \lambda_i \ (i = 1, 2, \dots n) \text{ are scalars,}$$

or $\quad 11 [1, 0, 0, \dots 0] + \lambda_2 [0, 1, 0, \dots, 0] + \dots + \lambda_n [0, 0, \dots, 1] = 0$

or $\quad [\lambda_1, \lambda_2, \lambda_3, \dots, \lambda_n] = [0, 0, 0, \dots, 0].$

This implies that $\lambda_i = 0$ for all $i = 1, 2, \dots n$.

Thus, the n vectors e_i, i = 1, 2, ... ,n, are linearly independent.

Let x be any other vector of R^n different from the vectors

e_i, i = 1, 2, ... ,n.

This can be written as

$$\begin{aligned} x &= [x_1, x_2, .. x_k, ... ,x_n] \\ &= x_1 e_1 + x_2 e_2 + ... + x_n e_i \\ &= \sum_{i=1}^{n} x_i e_i \end{aligned}$$

This happens to be a linear combination of the n unit vectors e_i; i = 1, 2, ... ,n. Since x is any vector of R^n different from the n unit vectors, therefore, every vector of R^n can be expressed as a linear combination of these vectors.

Hence {e, i = 1, 2, ... ,n} forms a basis for R^n.

Definition 3:

(Standard basis). The set of vectors e_i, i = 1, 2, ... n, is called the *Standard basis* for R^n.

Theorem 2:

(Replacement Theorem). *Let $\{a_1, a_2, ... ,a_n\}$ be any basis set for R^n and b ≠ 0 any other vector in R^n. If b is expressible as*

$$b = \sum_{i=1}^{n} \delta_i a_i \text{ where } d_i \neq 0, \text{ for some } i,$$

then the vector ak (for which $\delta_k \neq 0$) can be replaced by the vector b. The new set of n vectors $a_1, a_2, ... , a_{k-1}, a_{k+1}, ... ,a_n$, b is also a basis set for R^n.

Proof:

Let $\delta_k \neq 0$. Consider the new set of vectors

$$a_1, a_2, ... , a_{k-1}, a_{k+1}, ... ,a_n, b.$$

To prove that this set is a basis for R^n, we must show

(i) $a_1, a_2, ... , a_{k-1}, a_{k+1}, a_n$, b are linearly independent, and

(ii) any vector of R^n can be expressed as a linear combination of the vectors $a_1, a_2, ... , a_{k-1}, a_{k+1}, ... ,a^n$, b.

Let us suppose the set $\{a_1, a_2, \dots, a_n, \dots b\}$ 10 net a linearly independent set. Then the equation $\sum_{i=j}^{n} \lambda_i a_i + \lambda_b = 0$

will hold, even with some non-zero scalars λi $[i = 1, 2, \dots, k-1, k+1 \dots n]$

now $\lambda = 0$ implies

$$\sum_{i=1}^{n} \lambda_i a_i = 0 \text{ with bow } \lambda_i \neq 0.$$

This contradicts the feat third the set $[a_1, a_2, \dots a_{ks}, a_{k+1} \dots a_u]$ is linearly independent set,

Thus $l \neq 0$ amul wo have

$$\sum_{i=1}^{n} \lambda_i a_i + \lambda \sum_{i=1}^{n} \delta_i a_i = 0 \left[\text{since } b = \sum_{i=1}^{n} \delta_i a_i \right]$$

$$\sum_{i=1 i \neq k}^{n} (\lambda i + \lambda \delta i) ai + \lambda a\, \delta ak = 0$$

Since $l \neq 01$ $d_u \neq c$ the vector the set of vectors $a_1, a_2 \dots a_n$ is linearly dependent. This centralised the hypothesis.

Thus $\{a_1, a_2 \dots a_{k-1}, a_{k+1}, \dots a_n, b]$ is a linearly in dependent set

Now let x be any vector in R^n other thou the given basis vectors $a_1, a_2 \dots a_n$ then we can express x as

$$n = \sum_{i=1}^{n} \lambda_i a_i \text{ where } \lambda_i \ [i = 1, 2, \dots, n] \text{ are scalars.}$$

$$\text{Now since } b = \sum_{i=1}^{n} \delta_i a_i = \sum_{i=1 i=k}^{n} \delta_i a_i + \delta_k a \quad , \delta_1 \neq 0$$

$$\text{we have } a_u = \frac{1}{\delta u} \left[b - \sum_{i=1, i \neq k}^{n} \delta_i a_i \right]$$

$$= \frac{b}{\delta k} - \sum_{i=1, i \neq k}^{n} \frac{bi}{\delta k} a_i$$

$$x = \sum_{i=1, i \neq k}^{n} \lambda_i a_i + \lambda_k \left[\frac{b}{\delta_k} - \sum_{i=1, i \neq k}^{n} \frac{\delta_i}{\delta_k} a_i \right]$$

$$= \sum_{i=1, i \neq k}^{n} \left(\lambda_1 - \frac{\lambda_k}{\delta_k} \delta_i \right) a_i + \frac{\lambda_k}{\delta_k} b.$$

Thus, x can be expressed as a linear combination of the vectors a_1, a_2, ... ,a_{k-1}, a_{k+1}, a_n, b.

Hence the set $\{a_1, a_2, \ldots, a_{k-1}, a_{k+1}, \ldots, a_n, b\}$ forms a basis for R_n.

1. Point Sets. *Point sets are sets whose elements are points or vectors in E^n or R^n (n-dimensional Euclidean space).*

For example,

(i) A linear equation in two variables x_1, x_2, *i.e.*, $a_1 x_1 + a_2 x_2 = b$ represents a line in two dimensions. This line may be considered as a set of those points (x_1, x_2) which satisfy $a_1 x_1 + a_2 x_2 = b$. This set of points can be written as

$$S_1 = \{(x_1, x_2): a_1 x_1 + a_2 x_2 = b\}.$$

(ii) Consider the set of points lying inside a circle of unit radius with centre at the origin, in two dimensional space (E^2). Obviously the points (x_1, x_2) of this set satisfy the inequality $x_1^2 + x_2^2 < 1$.

This set of points can be written as

$$S_2 = \{(x_1, x_2): x_1^2 + x_2^2 < 1\}.$$

These sets may contain either a finite or infinite number of elements. Usually, however, they will contain an infinite number of elements. Further, we shall always assume that there is at least one element in a set unless otherwise stated.

2. Hypersphere. *A hypersphere in En with centre at a and radius $\varepsilon > 0$ is defined to be the set of points*

$$X = \{x: |x - a| = \varepsilon\}$$

i.e., the equation of a hypersphere in E^n is

$$(x_1 - a_1)^2 + (x_2 - a_2)^2 + \ldots + (x_n - a_n)^2 = \varepsilon^2$$

where $a = (a_1, a_2, \ldots a_n)$,

$x = (x_1, x_2, \ldots x_n)$,

which represents a circle in E^2 and sphere in E^3.

The set of points inside the hypersphere in E^3.

The set of points inside the hypersphere is the set

$$X = \{x : |x - a| < \varepsilon \}.$$

3. **An $\in$-neighbourhood.** *An neighbourhood about the point a is defined as the set of points lying inside the hypersphere with centre at a and radius e > 0.*

   i.e., the Î-neighbourhood about the point a is the set of points

   $$X = \{x : |x - a| < \varepsilon\}.$$

4. **An Interior Point.** *A point a is an interior point of the set S if there exists an ε-neighbourhood about a which contains only points of the set S.*

   An interior point of S must be an element of S.

5. **Boundary Point.** *A point a is a boundary point of the set S if every ε-neighbourhood about a (e > 0 may be, however, small) contains points which are in the set and the point which are not in the set.*

   A boundary point of S does not have to be an element of S.

CONVEX POLYHEDRON

The set of all convex combinations of finite number of points is called the convex polyhedron generated by these points.

Alternatively, if the set X consists of a finite number of points, the convex hull of X is called a convex polyhedron with vertices at these points.

Example:

The set of the area of a triangle is a convex polyhedron of the set of its vertices.

IMPORTANT THEOREMS

Theorem 1:

A hyperplane is a convex set.

Proof:

Consider the hyperplane

$$X = \{x : c\,x = z\}.$$

Let x_1 and x_2 be any two points in the hyperplane X.

$\therefore c\, x_1 = z$ and $c\, x_2 = z$.

If $x_3 = \lambda x_1 + (1 - \lambda)\, x_2,\ 0 \le \lambda \le 1$

then $cx_3 = \lambda c\, x_1 + (1 - \lambda)\, cx_2,$

$= \lambda z + (1 - \lambda)\, z = z$

which implies that

$x_3 = \lambda x_1 + (1 - \lambda)\, x_2$ is also a point in the hyperplane X.

Hence by definition, the hyperplane X is a convex set.

Theorem 2:

The closed half spaces $H_1 = \{x : c\,x \ge z\}$ and $H_2 = \{x : c\,x \le z\}$ are convex sets.

Proof:

Let x_1 and x_2 be any two points of H_1. Then $cx_1 \ge z$, $cx_2 \ge z$.

If $0 \le 1 \le 1$, then $c\,[\lambda x_1 + (1 - \lambda)\, x_2] = \lambda\, cx_1 + (1 - \lambda)\, cx_2$

$\ge \lambda z + (1 - \lambda)\, z = z.$

Hence $x_1, x_2 \in H_1$ and $0 \le 1 \le 1$

$\Rightarrow \quad \lambda x_1 + (1 - \lambda)\, x_2 \in H_1.$

So H_1 is a convex set.

Similarly, if $x_1, x_2 \in H_2$, $0 \le \lambda \le 1$, then replacing the inequality sign $\ge$ by $\le$ in above, we get st$x_1 + (1 - \lambda)\, x_2 \in H_2$.

So H_2 is also a convex set.

Corollary:

The open half spaces: $\{x : cx > z)$ and $\{x : cx < z\}$ are convex sets.

Theorem 3:

(a) *Intersection of two convex sets is also a convex set.*

Proof:

Consider tow convex sets X_1 and X_2. Let X_3 be the intersection of sets X_1 and X_2, *i.e.,* $X_3 = X_1 \cap X_2$.

Now $\quad x_1 \in X_1 \cap X_1$

$\Rightarrow \quad x_1 \in X_1$ and $x_1 \in X_2$

$$x_2 \in X_1 \cap X_2$$

$\Rightarrow$ $x_2 \in X_1$ and $x_2 \in X_2$.

Since X_1 and X_2 are convex sets,

$\therefore$ $x_1, x_2 \in X_1 \Rightarrow \lambda x_1 + (1 - \lambda) x_2 \in X_1$ $\quad 0 \le 1 \le 1$

and $x_1, x_2 \in X_2 \Rightarrow \lambda x_1 + (1 - \lambda) x_2 \in X_2$ $\quad 0 \le 1 \le 1$

Thus $\lambda x_1 + (1 - \lambda) x_2 \in X_1$ and $\lambda x_1 + (1 - \lambda) x_2 \in X_2$

$\Rightarrow \lambda x_1 + (1 - \lambda) x_2 \in X_1 \cap X_2$ $\quad 0 \le 1 \le 1.$

Hence by definition, $X_3 = X_1 \cap X_2$ is a convex set.

(b) *Intersection of any finite number of convex sets is also a convex set.*

Theorem 4:

The set of all convex combinations of a finite number of points x_1, x_2, ... ,x_n is a convex set.

Proof:

Let X be the set of all convex combinations of a finite number of point.

i.e., $$X = \left\{x : x = \sum_{i=1}^{n} \lambda_i x_i, \sum_{i=1}^{n} \lambda_i = 1, \lambda_i \ge 0\right\}.$$

Let $u, v \in X$

$$\therefore u = \sum_{i=1}^{n} a_i x_i, \sum_{i=1}^{n} a_i = 1,\ a_i \ge 0$$

and $$v = \sum_{i=1}^{n} b_i x_i \sum_{i=1}^{n} b_i = 1,\ b_i \ge 0.$$

Consider $w = \lambda u + (1 - \lambda) v,\ 0 \le \lambda \le 1$

$$\therefore w = 1 \sum_{i=1}^{n} a_i x_i + (1 - \lambda) \sum_{i=1}^{n} b_i x_u$$

$$= \sum_{i=1}^{n} \{\lambda a_i + (1 - \lambda) b_i\} x_i$$

$$- \sum_{i=1}^{n} c_i x_i \text{ where } c_i = \lambda a_i + (1 - \lambda)\ b_i$$

Now $\sum_{i=1}^{n} c_i = \sum_{i=1}^{n} \{\lambda a_i + (1-\lambda) b_i\}$

$= 1 \sum_{i=1}^{n} a_i + (1-\lambda) \sum_{i=1}^{n} b_i = \lambda.1 + (1 - \lambda).\ 1 = 1.$

Also $c_i = \lambda a_i + (1 - \lambda)\ b_i \geq 0$, V i.

Hence $w = \sum_{i=1}^{n} c_i x_i$ is a convex combination of $x_1, x_2, \ldots, x_n$.

i.e., $w = \lambda u + (1 - \lambda)\ v \in X, 0 \leq l \leq 1.$

Hence by definition X is a convex set.

Theorem 5:

Let S and T be two convex sets in E^n, then for any scalars α, β, αS + βT is also convex.

Proof:

Let x, y be two points of $\alpha S + \beta T$.

Then $x = \alpha u_1 + \beta v_1$ and $y = \alpha u_2 + \alpha v_2$ ...(1)

where $u_1, u_2 \in S$ and $v_1, v_2 \in T$.

For any scalar l, $0 \leq l \leq 1$, we have

$\lambda x + (1 - \lambda)\ y = \lambda\ (\alpha u_1 + \beta v_1) + (1 - \lambda)\ (\alpha u_2 + \beta v_2)$

$= \alpha\ [\lambda u_1 + (1 - \lambda)\ u_2] + \beta\ [\lambda v_1 + (1 - \lambda)\ v_2]$ ...(2)

But S and T are convex sets,

$u_1, u_2 \in S$

$\Rightarrow \lambda u_1 + (1 - \lambda)\ u_2 \in S, 0 \leq \lambda \leq 1$ ...(3)

and $v_1, v_2 \in T$

$\Rightarrow \lambda v_1 + (1 - \lambda)\ v_2 \in T, 0 \leq \lambda \leq 1.$ ...(4)

Now from (2), (3) and (4) we have

$\lambda x + (1 - {}_l)\ y \in \alpha S + \beta T, 0 \leq l \leq 1.$

Thus $x, y \in \alpha S + \alpha T$

$\Rightarrow [x : y] \subset \alpha S + \beta T.$

Hence $\alpha S + \beta T$ is a convex set.

Corollary:

If S and T be two convex sets in E^n, then S + T and S − T are also convex sets.

Theorem 6:

A set C is convex iff every convex linear combination of points in C, also belongs to C.

Proof:

Let every convex linear combination of points in C belong to C.

Then, in particular convex linear combination of every two points in C also belong to C.

Hence C is a convex set.

Conversely let C be a convex set. Then to prove that convex linear combination of any number of points in C also belong to C.

We shall use the induction principle.

Since C is convex, the convex linear combination of two points in C belongs to C. Thus the result is true for n = 2.

Now suppose the convex linear combination of any n points in C, belongs to C.

Let $x_1, x_2, \ldots, x_n$ be any n points in C. Then by assumption, $\lambda_1 x_1 + \lambda_2 x_2 + \ldots + \lambda_n x_n \in C$ where $\lambda_i \geq 0$, $\Sigma\lambda_i = 1$.

Let $x = \mu_1 x_1 + \mu_2 x_2 + \ldots + \mu_n x_n + \mu_{n+1} x_{n+1}$ ...(1)

i.e., x is a convex linear combination of (n + 1) points of C, where

$\mu_i \geq 0$ and $\sum_{i=1}^{n+1} \mu_i = 1$.

Now we shall show that $x \in C$.

If $m_{n+1} = 0$ then x becomes a convex linear combination of $x_1, x_2, \ldots, x_n$ which by assumption, belongs to C and hence the result holds in this case. Also if $m_{n+1} = 1$, then the result is trivially true.

Let m_{n+1} be neither 0 nor 1. Then

$$x = \frac{(\mu_1 + \mu_2 + \ldots + \mu_n)}{(\mu_1 + \mu_2 + \ldots + \mu_n)}(\mu_1 x_1 + \mu_2 x_2 + \ldots + \mu_n x_n) + \mu_{n+1} x_{n+1}$$

...(2)

$$= \left(\sum_{i=1}^{n}\mu_i\right)\left(\sum_{i=1}^{n}a_i x_i\right) + \mu_{n+1}x_{n+1},$$

where $a_i = \dfrac{\mu_i}{\mu_i + \mu_2 + \ldots + \mu_n}$, i = 1, 2, ... ,n.

As each $\mu_i \geq 0$, we have $a_i \geq 0$

and also $\sum_{i=1}^{n} a_i = \sum_{i=1}^{n}\left(\dfrac{\mu_i}{\mu_1 + \mu_2 + \ldots + \mu_n}\right) = \dfrac{\Sigma\mu_i}{\Sigma\mu_i} = 1.$

Thus $\sum_{i=1}^{n} a_i x_i = y$ (say) is also a convex linear combination of x_1, x_2, ... ,x_n. So y belongs to C, by assumption.

Now $x = (\mu_1 + \mu_2 + \ldots + \mu_n)\, y + \mu_{n+1}\, x_{n+1}$ ...(3)

But $(\mu_1 + \mu_2 + \ldots + \mu_n) \geq 0,\ \mu_{n+1} \geq 0$

and $(\mu_1 + \mu_2 + \ldots + \mu_n) + \mu_{n+1} = 1.$

It follows that x is a convex combination of two points y and x_{n+1} of C. So x belongs to C.

Hence convex linear combination of (n + 1) points of C also belongs to C. So by induction principle the result is true.

Theorem. 7:

The set of all feasible solution (if not empty) of a L.P.P. is a convex set.

Proof:

Let X be the set of all feasible solution of a L.P.P.

Case I.

If the set X has only element, then X is convex set. Hence the theorem is true in this case.

Case II.

If the set X has at least two elements.

Let x_1 and x_2 be any two distinct elements in X.

$\therefore \quad A x_1 = b, x_1 \geq 0.$

and $\quad A x_2 = b, x_2 \geq 0.$

If $\quad x_3 = \lambda x_1 + (1 - \lambda) x_2, 0 \leq \lambda \leq 1$

then
$$Ax_3 = A\lambda x_1 + (1 - \lambda) Ax_2$$
$$= \lambda b + (1 - \lambda) b = b.$$

Also since $x_1 \geq 0, 2 \geq 0, \lambda \geq \lambda \geq 0$, as $0 \leq \lambda \leq 1$

$$\therefore \quad x_3 = \lambda x_1 + (1 - \lambda) x_2 \geq 0$$

i.e., x_3 satisfies (1). Thus $x_3 = \lambda x_1 + (1 - \lambda) x_2$ is also a F.S. and so belongs to set X.

But x_3 is a convex combination of any two distinct points, x_1 and x_2 in X.

Hence by definition the set X is a convex set.

Note: Since the convex combinations of two points are infinite in number so from the above theorem we conclude that:

If a given L.P.P. has two feasible solutions, then it has infinite number of feasible solutions.

Theorem 8:

Every basic feasible solution of the system $Ax = b$, $x \geq 0$ is an extreme point of the convex set of feasible solutions and conversely.

Proof:

To prove that every B.F.S. is an extreme point of the convex set of all feasible solution.

Let x be a B.F.S. of Ax = b which is a n-component vector containing both zero (non-basic) and non-zero (basic) variables. Let x_B and B be the vector of m basic variables and the matrix of vectors associated to basic variables in the B.F.S. x respectively, then

$$x = [x_B, 0], \qquad \text{...(1)}$$

where 0 is a null vector of (n – m) components,

and $\quad Ax = b \Rightarrow B. x_B = b. \qquad$ ...(2)

Now we have to prove that x is an extreme point.

We shall prove this by using contradiction.

Suppose that x is not an extreme point. If X is the convex set of all feasible solutions of Ax = b, then $x \in X$.

If x is not an extreme point then there exist two distinct points x_1 and x_2 in X such that

$$x = \lambda x_1 + (1 - \lambda) x_2, 0 < \lambda < 1 \quad ...(3)$$

But x_1 and x_2 can be expressed as

$$x_1 = [u_1, v_1] \text{ and } x_2 = [u_2, v_2] \quad ...(4)$$

where u_1 and u_2 are vectors of m components of x_1 and x_2 respectively and v_1, v_2 are (n – m) component vectors.

Substituting the values of x and x_1, x_2 from (1) and (4) in (3), we have

$$[x_B, 0\} = \lambda [u_1, v_1] + (1 - \lambda) [u_2, v_2], = < \lambda < 1$$

$$= [\lambda u_1 + (1 - \lambda) u_2, \lambda v_1 + (1 - \lambda) v_2]$$

$$\therefore 0 = \lambda v_1 + (1 - \lambda) v_2, 0 < \lambda < 1 \quad ...(5)$$

Now $1 > \lambda > 0$, $1 - \lambda > 0$, and the components of v_1 and v_2 are ≥ 0.

The relation (5) can only be satisfied when $v_1 = 0$ and $v_2 = 0$

$$\therefore \quad x_1 = [u_1, 0], x_2 = [u_2, 0].$$

Since x_1 and x_2 are in X, therefore from (2) we have

$$A x_1 = B u_1 = b$$

and $$A x_2 = B u_2 = b$$

i.e., $$b = B x_B = B u_1 = B u_2$$

which gives $x_B = u_1 = u_2$

$$\therefore \quad x = x_1 = x_2$$

which is contradiction to the assumption that $x_1 \neq x_2$

i.e., x cannot be expressed as a convex combination of any two distinct points in the set of all feasible solutions. Hence x must be an extreme point.

Converse. *To prove that every extreme point of the convex set of feasible solutions is a B.F.S.*

Let $x = [x_1, x_2, ... ,x_n]$ be an extreme point. Now in order to prove that x is a B.F.S. we shall prove that the vectors associated with the positive elements of x are L.I. Suppose that k-components (variables) in x are non-zero and (n – k) components are zero.

We can assume these components as the first k components of x.

$$\therefore \sum_{i=1}^{n} x_i \alpha_i = b, \; x_i > 0, \; i = 1, 2, \ldots, k. \qquad \ldots(6)$$

where α_i is the column vector in A associated to the ith variables in x.

If possible, let the column vectors $a_1, a_2, \ldots, a_k$ of matrix A be L.D. Then there exist some scalars l_i $(i = 1, 2, \ldots, k)$ with at least one of them

n o n

zero, s.t.,

$$\sum_{i=1}^{n} \lambda_i \alpha_i = 0 \qquad \ldots(7)$$

From (6) and (7), for some arbitrary $\delta > 0$, we have

$$\sum_{i=1}^{k} x_i \alpha_i \pm \delta \sum_{i=1}^{k} \lambda_i \alpha_i = b$$

$$\text{or } \sum_{i=1}^{n} (x_i \pm \delta \lambda_i) \alpha_i = b$$

from which it is obvious that the two points

(n – k) in number

$$x_1^* = [x_1 + \delta\lambda_1, x_2 + \delta\lambda_2, \ldots, x_k + \delta\lambda_k, 0, 0, \ldots, 0]$$

$$x_2^* = [x_1 - \delta\lambda_1, x_2 - \delta\lambda_2, \ldots, x_k - \delta\lambda_k, 0, 0, \ldots, 0]$$

satisfy A x = b.

Also since $x_i > 0$, therefore, taking δ s.t.

$$0 < d < \min_{i} \left\{ \frac{x_i}{|\lambda_i|} \right\}, \; \lambda_i \neq 0, \; i = 1, 2, \ldots, k.$$

We conclude that first k components of x_1^* and x_2^* are always positive. But the remaining components of x_1^* and x_2^* are zero, which follows that x_1^* and x_2^* are feasible solutions different from x.

Now $x_1^* + x_2^* = 2\,[x_1, x_2, \ldots, x_k, 0, 0, \ldots, 0]$

$$\text{or} \quad \frac{1}{2} x_1^* + \frac{1}{2} x_2^* = [x_1, x_2, \ldots x_k, 0, 0, \ldots, 0] = x$$

or $\quad x = \lambda x_1^* + (1 - \lambda) x_2^*$ where $\lambda = \frac{1}{2}$

i.e., x can be expressed as a convex combination of two distinct feasible solutions x_1^* and x_2^*.

But this is a contradiction as x is an extreme point. Hence the vectors $a_1, a_2, \ldots a_k$ are L.I. Further, we know that at most m vectors of Em can be L.I. So $a_1, a_2, \ldots a_k$ cannot be more than m and hence the extreme point x will have almost m non-zero variables, *i.e.,* atleast (n – m) variables will be zero.

Thus x is a B.F.S. Hence, every extreme points of the convex set of feasible solutions is a B.F.S.

Cor 1 : *The extreme points of the convex set of feasible solutions are finite in number.*

From the above theorem, we conclude that there is only one extreme point for a given B.F.S. and vice versa. That is there is one-to-one correspondence between the extreme points and the B.F. solutions in the absence of degeneracy. Also in case of degeneracy corresponding to an extreme point with the number of non-zero variables less than m, we can form more than one degenerate B.F.S. Hence the number of extreme points of the feasible region is finite and it cannot exceed the number of its B.F. solutions.

Cor 2: *An extreme point can have almost m-positive x_i's where m is the number of constraints.*

Cor 3 : *In an extreme point, vectors associated to the positive x_i's are L.I.*

Theorem 9:

If the convex set of the feasible solutions of $Ax = b, x \geq 0$ is a convex polyhedron, then atleast one of the extreme points gives an optimal solution.

Proof:

In the Cor. 1 of last theorem we have proved that the extreme points of the convex set of feasible solutions of $Ax = b$, $x \geq 0$ are finite in number.

Let $x_1, x_2, \ldots, x_k$ be the extreme points of the set X of all the feasible solutions of $Ax = b, x \geq 0$. Let Z be the objective function which is to be maximized be given by

$$Z = c\,x.$$

If $x^* \in X$ is the optimal solutions, then

$$\text{Max. } Z = c\,x^*.$$

Now if x^* is an extreme point, them the theorem is proved.

Now if x^* is not an extreme point in X, then since X is convex polyhedron, therefore x^* can be expressed as a convex combination of the extreme points of X,

i.e., $$x^* = \lambda_1 x_1 + \lambda_2 x_2 + \ldots + \lambda_k x_k = \sum_{i=1}^{k} \lambda_i . x_i,$$

$$l_i \geq 0 \text{ and } \Sigma\lambda_i = 1$$

$$\therefore Z^* = cx^* = c(\lambda_1 x_1 + \lambda_2 x_2 + \ldots + \lambda_k x_k)$$

$$= (\lambda_1 cx_1 + \lambda_2 cx_2 + \ldots + \lambda_k cx_k)$$

If maximum of sx_i is cx_p, then

$$Z^* \leq (\lambda_1 + \lambda_2 + \ldots + \lambda_k).\ cx_p$$

or $$Z^* \leq cx_p.$$

But Z^* is the maximum value of Z. Therefore,

$$Z^* = cx_p \text{ or } cx^* = cx_p$$

i.e., $x^* = x_p$ (one of the extreme points).

Hence the optimal solution is attained at the extreme point.

This proves the theorem.

Theorem 10:

If the objective function of a L.P.P. assumes its optimal value at more than one extreme point, then every convex combination of these extreme points gives the optimal value of the objective function.

Proof:

Let us consider the L.P.P. as follows:

$$\text{Max. } Z = cx$$

s.t. $\quad Ax = b, x \geq 0.$

Let $x_1, x_2, \ldots x_k$ be the extreme points of the feasible region. If the objective function Z assumes its optimal value Z* at the extreme points $x_1, x_2, \ldots, x_p$, $(p \leq k)$ then

$$Z^* = cx_1 = cx_2 = \ldots = cx_p.$$

If x_0 is the convex combination of the extreme points $x_1, x_2, \ldots x_p$, then

$$x_0 = \lambda_1 x_1 + \lambda_2 x_2 + \ldots + \lambda_p x_p, \lambda_i \geq 0, \sum_{i=1}^{p} \lambda_i = 1$$

$$\therefore cx_0 = c\,[\lambda_1 x_1 + \lambda_2 x_2 + \ldots + \lambda_p x_p]$$

$$= \lambda_1 cx_1 + \lambda_2 cx_2 + \ldots + \lambda_p cx_p$$

$$= \lambda_1 Z^* + \lambda_2 Z^* + \ldots + \lambda_p Z^*$$

$$= (\lambda_1 + \lambda_2 + \ldots + \lambda_p) Z^* = Z^*, \qquad \because \sum_{i=1}^{p} \lambda_i = 1.$$

Hence the optimal value Z* is also attained at x_0 which is the convex combination of the extreme points at which optimal value occurs. Hence the theorem.

SOLVED EXAMPLES

Example 1:

Show that $C = \{(x_1, x_2) : 2x_1 + 3x_2 = 7\} \subset R^2$ is a convex set.

Solution:

Let $u, v \in C$, where $u = (u_1, u_2)$, $v = (v_1, v_2)$.

Then $2u_1 + 3u_2 = 7$ and $2v_1 + 3v_2 = 7$. ...(7)

If $w = (w_1, w_2)$ is a point on the line segment joining the points u and v, then

$$w = \lambda u + (1 - \lambda)\, v, \; 0 \leq 1 \leq 1$$

$$\Rightarrow (w_1, w_2) = \lambda\,(u_1, u_2) + (1 - \lambda)\,(v_1, v_2)$$

$$= (\lambda u_1 + (1 - \lambda)\, v_1, \lambda u_2 + (1 - \lambda)\, v_2)$$

$$w_1 = \lambda u_1 + (1 - \lambda)\, v_1 \text{ and } w_2 = \lambda u_2 + (1 - \lambda)\, v_2.$$

$$2w_1 + 3w_2 = 2\,[\lambda u_1 + (1 - \lambda)\, v_1] + 3\,[\lambda u_2 + (1 - \lambda)\, v_2]$$

$= \lambda\ [2u_1 + 3u_2] + (1 - \lambda)\ [2v_1 + 3v_2]$

$= \lambda.\ 7 + (1 - \lambda).\ 7 = 7$, using (1)

$\therefore w = (w_1, w_2) \in C.$

Hence the set C is a convex set.

Example 2(a):

Show that $S = \{(x_1, x_2, x_3) : 2x_1 - x_2 + x_3 \leq 4\} \subset R^3$, is a convex set.

Solution:

Let $x = (x_1, x_2, x_3)$ and $y = (y_1, y_2, y_3)$ be any two point of S. Then we have

$$2x_1 + - x_2 + x_3 \leq 4 \text{ and } 2y_1 - y_2 + y_3 \leq 4 \qquad ...(1)$$

If $w = (w_1, w_2, w_3)$ is a point on the line segment joining the points x and y, then

$w = \lambda x + (1 - \lambda)\ y,\ 0 \leq 1 \leq 1$

$\Rightarrow (w_1, w_2, w_3) = \lambda\ (x_1, x_2, x_3) + (1 - \lambda)\ (y_1, y_2, y_3)$

$\Rightarrow w_1 = \lambda x_1 + (1 - \lambda)\ y_1,\ w_2 = \lambda x_2 + (1 - \lambda)\ y_2,\ w_3$

$= \lambda x_3 + (1 - \lambda)\ y_3.$

Now $2w_1 - w_2 + w_3 = \lambda\ (2x_1 - x_2 = x_3) + (1 - \lambda)\ (2y_1 - y_2 + y_3)$

$\leq 4\lambda + 4\ (1 - \lambda)$, using (1)

$= 4$

$\therefore\ w = (w_1, w_2, w_3) \in S.$

Hence the set S is a convex set.

Example 2(b):

Examine convexity to the set

$$\{(x_1, x_2) \in R^2 : 4x_1 + 3x_2 \leq 6,\ x_1 + x_2 \geq 1\}$$

Solution:

Let $S = \{(x_1, x_2) \in R^2 : 4x_1 + 3x_2 \leq 6$ and $x_1 + x_2 \geq 1$

If $u = (x_1, x_2) \in S$ Then $4x_1 + 3x_2 < 6$ and $x_1 + x_2 \geq 1$

and if $v = (y_1, y_2) \in S$ then $4y_1 + 3y_2 \geq 6$ and $x_1 + y_2 \geq 1$.

If $w = (z_1, z_2)$ is a point on the line segment joining points u and v, then

$$w = \lambda u + (1 - \lambda) v \qquad 0 \le 1 \le 1$$

$$\Rightarrow (z_1, z_2) = \lambda (x_1, x_2) + (1 - \lambda) (y_1, y_2)$$

$$= (\lambda x_1 + (1 - \lambda) y_1, \lambda x_2 + (1 - \lambda) y_2)$$

$$\Rightarrow z_1 = \lambda x_1 + (1 - \lambda) y_1 \text{ and } z_2 = \lambda x_2 + (1 - \lambda) y_2$$

Now $4z_1 + 3z_2 = \lambda (4x_1 + 3x_2) + (1 - \lambda) (4y_1 + 3y_2) \le 6$*

$\because 0 \le \lambda \le 1, 0 \le 1 - \lambda \le 1, 4x_1 + 3x_2 \le 6$ and $4y_1 + 3y_2 \le 6.$

and $z_1 + z_2 = \lambda (x_1 + x_2) + (1 - \lambda) (y_1 + y_2) \ge 1$**

$\because 0 \le \lambda \le 1, 0 \le 1 - \lambda \le 1, x_1 + x_2 \ge 1, y_1 + y_2 \ge 1.$

$\because w = (z_1, z_2) \in S.$

Hence the set S is a convex set.

Note:

As the extreme case if $4x_1 + 3x_2 = 6$, $4y_1 + 3y_2 = 6$,

then $4z_1 + 3z_2 = \lambda. 6 + (1 - \lambda) 6 = 6$

and if $x_1 + x_2 = 1$, $y_1 + y_2 = 1$ then $z_1 + z_2 = \lambda. 1 + (1 - \lambda). 1 = 1.$

Aliter.:

Obviously S is the intersection of two half spaces, viz.,

$$H_1 = \{(x_1, x_2) : 4x_1 + 3x_2 \le 6\} \text{ and } H_2 = \{(x_1, x_2) : x_1 + x_2 \ge 1\}.$$

Since H_1 and H_2 are convex sets so $S = H_1 \cap H_2$ is also convex.

Example 2(c):

A hyperplane is given by the equation

$$3x_1 + 2x_2 + 4x_3 + 7x_4 = 8.$$

and in which half spaces do the following points (– 6, 1, 7, 2) and (1, 2, – 4, 1) lie.

Solution:

The given equation of the hyperplane is

$$3x_1 + 2x_2 + 4x_3 + 7x_4 = 8.$$

Substituting (– 6, 1, 7, 2 in the L.H.S., we get

L.H.S. = 3 (– 6) + 2. 1 + 4. 7 + 7. 2 = 26 > 8 = R.H.S.

∴ the point (– 6, 1, 7, 2) lies in the open half space,

$3x_1 + 2x_2 + 4x_3 + 7x_4 > 8.$

Ageing substituting (1, 2, – 4, 1) in the L.H.P., we get

L.H.S. = 3. 1 + 2. 2 + 4 (– 4) + 7. 1 = – 2 < 8 = R.H.S.

∴ the point (1, 2, – 4, 1) lies in the open half space

$3x_1 + 2x_2 + 4x_3 + 7x_4 < 8.$

Example 2(d):

Let A be an m × n matrix and b an m-vector, then show that $\{x \in R^n : Ax \leq b\}$ *is a convex set.*

Solution:

Let $A = [a_{ij}]_{m\times n}$, $x = (x_1, x_2, \ldots, x_n)$ and $b = (b_1, b_2, \ldots, b_m)$, then the set $S = \{x \in R^n : Ax \leq b\}$ can be written in m-inequalities:

$$a_{11} x_1 + a_{12} x_2 + \ldots + a_{1n} x_n \leq b_1$$

$$a_{21} x_1 + a_{22} x_2 + \ldots + a_{2n} x_n \leq b_2$$

$$\ldots \quad \ldots \quad \ldots \quad \ldots \quad \ldots$$

$$a_{m1} x_1 + a_{m2} x_2 + \ldots + a_{mn} x_n \leq b_n.$$

Thus, the set S is the intersection of m half spaces

$H_i = \{(x_1, x_2, \ldots, x_n) : a_{i1} x_1 + a_{i2} x_2 + \ldots + a_{in} x_n \leq b_i, i = 1, 2, \ldots, m\}.$

If follows that $S = \bigcap_{i=1}^{m} H_i$ is convex as each half space is convex.

Example 3:

Show that the set $S = \{x : x = (x_1, x_2, x_3), x_1^2 + x_2^2 + x_3^2 \leq 1\}$ *is a convex set.*

Solution:

Let x, y ∈ S, where $x = (x_1, x_2, x_3)$, $y = (y_1, y_2, y_3)$.

Then, by the given condition, we have

$$x_1^2 + x_2^2 + x_3^2 \leq 1 \qquad \ldots(1)$$

and $y_1^2 + y_2^2 + y_3^2 \leq 1.$ ...(2)

If $z = (z_1, z_2, z_3)$ is a point on the line segment joining the points x and y, then

$z = \lambda x + (1 - \lambda) y, 0 \leq \lambda \leq 1$

$\Rightarrow (z_1, z_2, z_3) = 1 (x_1, x_2, x_3) + (1 - \lambda) (y_1, y_2, y_3)$

$\Rightarrow z_1 = \lambda x_1 + (1 - \lambda) y_1, z_2 = \lambda x_2 + (1 - \lambda) y_2, z_3$

$= \lambda x_3 + (1 - \lambda) y_3.$

Now $z_1^2 + z_2^2 + z_3^2 = [\lambda x_1 + (1 - \lambda) y_1]^2 + [\lambda x_2 + (1 - \lambda) y_2]^2 + \lambda x_3 + (1 - \lambda) y_3]^2$

$= \lambda_2 (x_1^2 + x_2^2 + x_3^2) + (1 - \lambda)^2 (y_1^2 + y_2^2 + y_3^2) + 2\lambda (1 - \lambda) (x_1 y_1 + x_2 y_2 + x_3 y_3)$

$\leq \lambda_2 . 1 + (1 - \lambda)^2 . \lambda + 2_1 (1 - \lambda) (x_1 y_1 + x_2 y_2 + x_3 y_3)$, using (1) and (2)

By Lagrange's identity, we have

$(x_1^2 + x_2^2 + x_3^2) (y_1^2 + y_2^2 + y_3^2) - x_1 y_1 + x_2 y_2 + x_3 y_3)^2 = \Sigma (x_1 y_2 - x_2 y_1)^2 \geq 0$

$\therefore x_1 y_1 + x_2 y_2 + x_3 y_3 \leq \sqrt{(x_1^2 + x_2^2 + x_3^2)} \sqrt{(y_1^2 + y_2^2 + y_3^2)} \leq 1,$ using (1) and (2)

Thus $z_1^2 + z_2^2 + z_3^2 \leq \lambda_2 + \lambda_2 + (1 - \lambda)^2 + 2_1 . (1 - \lambda) = 1.$

$\therefore z = (z_1, z_2, z_3) \in S.$

Hence the set S is a convex set.

Example 4:

Find the extreme points, if any, the following sets:

(i) $S = \{(x, y) : x_2 + y_2 \leq 25\}$

(ii) $\{(x, y) : |x| \leq 1, |y| \leq 1\}$.

Solution:

(i) Draw the region of the set S. It represents the boundary and interior of the circle with centre at (0, 0) and radius 5.

Every point on its circumference is an extreme point.

(ii) We have $|x| \leq 1 \Rightarrow -1 \leq x \leq 1$ and $|y| \leq 1$

$\Rightarrow -1 \leq y \leq 1.$

Thus the set S represents the region of rectangle bounded by the lines $x = 1$, $x = -1$, $y = 1$, $y = -1$. The extreme points of this convex set are $(1, 1)$, $(-1, 1)$, $(-1, -1)$ and $(1, -1)$.

Example 5:

Is the union of two convex sets, necessarily a convex set?

Solution:

No, it is not necessary that union of two convex sets will be a convex set.

For example: $S_1 = \{(x, y) : x \geq 1\}$, $S_2 = \{(x, y) : x \geq 2\}$

and $T_1 = \{(x, y) : x \geq 1\}$, $T_2 = \{(x, y) : y \geq 1\}$.

Then $S_1 \cup S_2$ is again a convex set while $T_1 \cup T_2$ is not a convex set.

Example 6:

Find all the basic feasible solutions for the equations

$$2x_1 + 6x_2 + 2x_3 + x_4 = 3$$

$$6x_1 + 4x_2 + 4x_3 + 6x_4 = 2$$

$$x_i \geq 0$$

and determine the associated general convex combination of the extreme point solutions.

Solution:

In matrix form the given system of equations can be written as

$$A\,x = b$$

where $A = (\alpha_1, \alpha_2, \alpha_3, \alpha_4)$

$$\alpha_1 = \begin{bmatrix} 2 \\ 6 \end{bmatrix}, \alpha_2 = \begin{bmatrix} 6 \\ 4 \end{bmatrix}, \alpha_3 = \begin{bmatrix} 2 \\ 4 \end{bmatrix}, \alpha_4 = \begin{bmatrix} 1 \\ 6 \end{bmatrix}, x = \begin{bmatrix} x_1 \\ x_2 \\ x_3 \\ x_4 \end{bmatrix}, b = \begin{bmatrix} 3 \\ 2 \end{bmatrix}.$$

This problem can have at most ${}^4C_2 = 6$ basic solutions. Now the six sets of two vectors are

$$B_1 = [\alpha_1, \alpha_2] = \begin{bmatrix} 2 & 6 \\ 6 & 4 \end{bmatrix}, B_2 = [\alpha_1, \alpha_3] = \begin{bmatrix} 2 & 2 \\ 6 & 4 \end{bmatrix}$$

$$B_3 = [\alpha_1, \alpha_4] = \begin{bmatrix} 2 & 1 \\ 6 & 6 \end{bmatrix}, B_4 = [\alpha_2, \alpha_3] = \begin{bmatrix} 6 & 2 \\ 4 & 4 \end{bmatrix}$$

$$B_5 = [\alpha_2, \alpha_4] = \begin{bmatrix} 6 & 1 \\ 4 & 6 \end{bmatrix}, B_6 = [\alpha_3, \alpha_4] = \begin{bmatrix} 2 & 1 \\ 4 & 6 \end{bmatrix}.$$

Here $|B_1| = -28, |B_2| = -4,$

$|B_3| = 6, |B_4| = 16,$

$|B_5| = 32, |B_6| = 8.$

Since none of these is zero, therefore all these sets are L.I.

Hence all the six basic solutions exist.

If x_{Bi}, i = 1, 2, ... , 6 are the vectors of the basic variables associated to the sets B_i, i = 1, 2, ... , 6 respectively, then

$$\begin{bmatrix} x_1 \\ x_2 \end{bmatrix} = x_{B1} = B_1^{-1} b = -\frac{1}{28}\begin{bmatrix} 4 & -6 \\ -6 & 2 \end{bmatrix}\begin{bmatrix} 3 \\ 2 \end{bmatrix} = \begin{bmatrix} 0 \\ 1/2 \end{bmatrix}$$

$$\begin{bmatrix} x_1 \\ x_3 \end{bmatrix} = x_{B2} = B_2^{-1} b = -\frac{1}{4}\begin{bmatrix} 4 & -2 \\ -6 & 2 \end{bmatrix}\begin{bmatrix} 3 \\ 2 \end{bmatrix} = \begin{bmatrix} -2 \\ 7/2 \end{bmatrix}$$

$$\begin{bmatrix} x_1 \\ x_4 \end{bmatrix} = x_{B3} = B_3^{-1} b = \frac{1}{6}\begin{bmatrix} 6 & -1 \\ -6 & 2 \end{bmatrix}\begin{bmatrix} 3 \\ 2 \end{bmatrix} = \begin{bmatrix} 8/3 \\ -7/3 \end{bmatrix}$$

$$\begin{bmatrix} x_2 \\ x_3 \end{bmatrix} = x_{B4} = B_4^{-1} b = \frac{1}{16}\begin{bmatrix} 4 & -2 \\ -4 & 6 \end{bmatrix}\begin{bmatrix} 3 \\ 2 \end{bmatrix} = \begin{bmatrix} 1/2 \\ 0 \end{bmatrix}$$

$$\begin{bmatrix} x_2 \\ x_4 \end{bmatrix} = x_{B5} = B_5^{-1} b = \frac{1}{32}\begin{bmatrix} 6 & -1 \\ -4 & 6 \end{bmatrix}\begin{bmatrix} 3 \\ 2 \end{bmatrix} = \begin{bmatrix} 1/2 \\ 0 \end{bmatrix}$$

$$\begin{bmatrix} x_3 \\ x_4 \end{bmatrix} = x_{B6} = B_6^{-1} b = \frac{1}{8}\begin{bmatrix} 6 & -1 \\ -4 & 2 \end{bmatrix}\begin{bmatrix} 3 \\ 2 \end{bmatrix} = \begin{bmatrix} 2 \\ -1 \end{bmatrix}.$$

Thus it is obvious that out of these only three basic solutions are B.F.S. (in which variables) are non-negative). But the B.F.S.'s correspond to the extreme points. Hence the only three extreme point solutions are given by

$$x_1 = \left(0, \frac{1}{2,}, 0, 0\right),$$

$$x_2 = \left(0, \frac{1}{2}, 0, 0\right),$$

$$x_3 = \left(0, \frac{1}{2}, 0, 0\right).$$

Here $x_1 = x_2 = x_3$. Hence there is unique extreme point solution.

Note: To find the basic solution x_{B1} we can also proceed as follows. Putting $x_3 = 0$, $x_4 = 0$ in the given eqns., we get $2x_1 + 6x_2 = 3$ and $6x_1 + 4x_2 = 2$ solving $x_1 = 0$, $x_2 = 1/2$, etc.

Example 7:

Show that the set $S = \{x : x = (x_1, x_2, x_3), x_1^2 + x_2^2 + x_3^2 \le 1\}$ *is a convex set.*

Solution:

Let $x, y \in S$, where $x = (x_1, x_2, x_3)$, $y = (y_1, y_2, y_3)$.

Then, by the given condition, we have

$$x_1^2 + x_2^2 + x_3^2 \le 1 \qquad ...(1)$$

and $$y_1^2 + y_2^2 + y_3^2 \le 1. \qquad ...(2)$$

If $z = (z_1, z_2, z_3)$ is a point on the line segment joining the points x and y, then

$$z = \lambda x + (1 - \lambda) y, \ 0 \le \lambda \le 1$$

$$\Rightarrow (z_1, z_2, z_3) = 1 (x_1, x_2, x_3) + (1 - \lambda) (y_1, y_2, y_3)$$

$$\Rightarrow z_1 = \lambda x_1 + (1 - \lambda) y_1, z_2 = \lambda x_2 + (1 - \lambda) y_2, z_3$$

$$= \lambda x_3 + (1 - \lambda) y_3.$$

Now $z_1^2 + z_2^2 + z_3^2 = [\lambda x_1 + (1 - \lambda) y_1]^2 + [\lambda x_2 + (1 - \lambda) y_2]^2 + \lambda x_3 + (1 - \lambda) y_3]^2$

$= \lambda_2 (x_1^2 + x_2^2 + x_3^2) + (1 - \lambda)^2 (y_1^2 + y_2^2 + y_3^2) + 2\lambda (1 - \lambda) (x_1 y_1 + x_2 y_2 + x_3 y_3)$

$\le \lambda_2 . 1 + (1 - \lambda)^2 . \lambda + 2_1 (1 - \lambda) (x_1 y_1 + x_2 y_2 + x_3 y_3)$, using (1) and (2)

By Lagrange's identity, we have

$$(x_1^2 + x_2^2 + x_3^2)(y_1^2 + y_2^2 + y_3^2) - x_1 y_1 + x_2 y_2 + x_3 y_3)^2$$

$$= \Sigma (x_1 y_2 - x_2 y_1)^2 \ge 0$$

$\therefore x_1 y_1 + x_2 y_2 + x_3 y_3 \le \sqrt{(x_1^2 + x_2^2 + x_3^2)} \sqrt{(y_1^2 + y_2^2 + y_3^2)} \le 1$, using (1) and (2)

Thus $z_1^2 + z_2^2 + z_3^2 \leq \lambda_2 + \lambda_2 + (1 - \lambda)^2 + 2_1 . (1 - \lambda) = 1.$

$\therefore z = (z_1, z_2, z_3) \in S.$

Hence the set S is a convex set.

Example 8:

For any points $x, y \in R^n$, show that the line segment $[x : y]$ is a convex set.

Solution:

Let $u, v \in [x : y]$. Then

$$u = \lambda_1 x + (1 - \lambda_1) y, \; 0 \leq \lambda_1 \leq 1$$

and $$v = \lambda_2 x + (1 - \lambda_2) y, \; 0 \leq \lambda_2 \leq 1.$$

Now let w be a point on the line segment joining the points u and v.

Then $w = \lambda u + (1 - \lambda) v, \; 0 \leq \lambda \leq 1.$

$$= \lambda [\lambda_1 x + (1 - \lambda_1) y] + (1 - \lambda) [\lambda_2 x + (1 - \lambda_2) y]$$

$$= [\lambda\lambda_1 + (1 - \lambda) \lambda_2] x + [\lambda (1 - \lambda_1) + (1 - \lambda) (1 - \lambda_2)] y.$$

If we take $\mu = \lambda\lambda_1 + (1 - \lambda) \lambda_2$, then

$$1 - \mu = 1 - \lambda\lambda_1 - (1 - \lambda) \lambda_2 = \lambda + (1 - \lambda) - \lambda\lambda_1 - (1 - \lambda) \lambda_2$$

$$= \lambda (1 - \lambda_1) + (1 - \lambda) (1 - \lambda_2).$$

Since, $0 \leq \lambda_1 \leq 1, \; 0 \leq \lambda_2 \leq 1$

$\Rightarrow 0 \leq \lambda\lambda_1 + (1 - \lambda) \lambda_2 \leq 1,$

$$w = \mu x + (1 - \mu) y, \; 0 \leq \mu \leq 1$$

$\therefore w \in [x : y].$

Hence the set $[x : y]$ is a convex set.

Example 9:

For any points $x, y \in R^n$, show that the line segment $[x : y]$ is a convex set.

Solution:

Let $u, v \in [x : y]$. Then

$$u = \lambda_1 x + (1 - \lambda_1) y, \; 0 \leq \lambda_1 \leq 1$$

and $$v = \lambda_2 x + (1 - \lambda_2) y, \; 0 \leq \lambda_2 \leq 1.$$

Now let w be a point on the line segment joining the points u and v.

Then $w = \lambda u + (1 - \lambda) v, 0 \le \lambda \le 1$.

$= \lambda [\lambda_1 x + (1 - \lambda_1) y] + (1 - \lambda) [\lambda_2 x + (1 - \lambda_2) y]$

$= [\lambda\lambda_1 + (1 - \lambda) \lambda_2] x + [\lambda (1 - \lambda_1) + (1 - \lambda) (1 - \lambda_2)] y.$

If we take $\mu = \lambda\lambda_1 + (1 - \lambda) \lambda_2$, then

$1 - \mu = 1 - \lambda\lambda_1 - (1 - \lambda) \lambda_2 = \lambda + (1 - \lambda) - \lambda\lambda_1 - (1 - \lambda) \lambda_2$

$= \lambda (1 - \lambda_1) + (1 - \lambda) (1 - \lambda_2).$

Since, $0 \le \lambda_1 \le 1, 0 \le \lambda_2 \le 1$

$\Rightarrow 0 \le \lambda\lambda_1 + (1 - \lambda) \lambda_2 \le 1,$

$w = \mu x + (1 - \mu) y, 0 \le \mu \le 1$

$\therefore w \in [x : y]$. Hence the set $[x : y]$ is a convex set.

EXERCISES

1. If S_1 and S_2 be two non-empty disjoint sets and S be a set such that if $x_1 \in S_1$ and $x_1 \in S_2$ then $x_1 - x_2 \in S$. Show that S is also a convex set and does not contain the origin.
2. Show that the set of all the internal points of a convex set S is a convex set.
3. Determine whether the vector [7, 0] is a convex combination of the vectors [6, 3], [9, – 6], [1, 2], [1, –1].
4. Give an example of a convex set whose every boundary point is an extreme point.
5. Can there be any convex set without any extreme point? Prove that an extreme point of a convex set is a boundary point of the set.
6. Is the set $S = \{x : x \in E^m, |x| = 1\}$ convex?
7. A and B are two convex sets in R^n, C is a set in R^n defined as

   $C = \{z \in R^n : z = x + y,$

   $x \in A, y \in B\}.$

   Examine convexity of C.
8. Express (2, 1), $\left(0, \frac{3}{2}\right)$, if possible, as a convex combinations of (1, 1) and (–1, 2).

9. What is meant by convex combination of vectors? Prove that the set of all convex combinations of linearly independent vectors is a convex set.

10. Prove that the objective function of a L.P.P. assumes minimum value at an extreme point of the convex set X generated by the set of all feasible solutions.

11. Prove that the collection of feasible solutions of a L.P.P. constitutes a convex set, whose extreme points correspond to B.F. solutions.

12. If a set of $k \le m$ vectors $P_1, P_2, \ldots, P_k \in A$ can be found that are linearly independent and such that $x_1 P_1 + x_2 P_2 + \ldots + x_k P_k = b$, and all $x_i \ge 0$, then show that the point $x = (x_1, x_2, \ldots x_k, 0, 0 \ldots, 0)$ is an extreme point of the convex set of feasible solution of $Ax = b$.

13. Explain the procedure of generating extreme point solutions (Analytical method) to a linear programming problem pointing out the assumptions made, if any.

14. Which of the following sets are convex?

    (i) $A = \{(x_1, x_2) : x_1 x_2 \le 1, x_1 \ge 0, x_2 \ge 0\}$

    (ii) $A = \{(x_1, x_2) : x_2^2 - 3 \ge x_1^2, x_1 \ge 0, x_2 \ge 0\}$.

15. Prove that the set $\{(x_1, x_2) : x_1^2 + x_2^2 \le 4\}$ is a convex set.

16. Define a convex set. Show that the set $S = \{(x_1, x_2) : 3x_1^2 + 2x_2^2 \le 6\}$ is convex.

17. Given two planes $a_1x + b_1y + c_1z + d_1 = 0$, $a_2x + b_2y + c_2z + d_2 = 0$ in R^3, prove that their intersection is a convex set but their union is not.

18. Show that $S = \{(x_1, x_2, x_3) : 2x_1 - x_2 + x_3 \le 4, x_1 + 2x_2 - x_3 \le 1\}$ is a convex set.

5

Transportation Table and Their Properties

INTRODUCTION

Definition (Loop): In a Transportation table, an order set of four or more cells is said to form a loop if : (i) any two adjacent cells in the ordered set lie either in the same row or in the same column and (ii) any three or more adjacent cells in the ordered set do not lie in the same row or the same column the first call of the set is considered to follow the last in the set.

If we join the cells of a loop by horizontal and vertical segments we get a closed path satisfying : (i) and (ii) condition. Let us denote by (i, j) the (i, j)th cell of a transportation table. Then, it can be seen from the graphical illustration in the fig. given below: that the set

$L = \{(2, 1), (4, 1), (4, 4), (1, 4), (1, 2), (2, 2)\}$

forms at loop while the set

$L' = \{(3, 2), (3, 5), (2, 5), (2, 4), (2, 3), (1, 3), (1, 2)\}$

does not form a loop.

Definition: 2. (Set Containing a Loop) : A set x of cells of a transportation table is said to CONTAIN A LOOP if a call of x or of a subset of x can be ordered (sequenced) so as to form a loop.

Where $x = [x_{12} \ldots, x_{1n}, x_{21} \ldots, x_{2n} \ldots, x_{2n} \ldots, x_{m1} \ldots x_{mn}]$ C is the cost vector $b = \{a_1 \ldots a_m, b_1 \ldots b_n\}$ and A is an $(m + n) \times mn$ real matrix containing the coefficient of constraints.

The reader should note that the elements of A are either 0 or +1 thus the general L.P.P can be reduced to be a T.P. if

(i) The a_{ij}'s are restricted to the value of 0 and + 1, and

(ii) The units among the constraints are homogeneous for a T.P. involving 2 origins and 3 destination (m = 2, n = 3) the matrix A is given by

$$A = \left[\begin{array}{ccc|ccc} 1 & 1 & 1 & 0 & 0 & 0 \\ 0 & 0 & 0 & 1 & 1 & 1 \\ \hline 1 & 0 & 0 & 1 & 0 & 0 \\ 0 & 1 & 0 & 0 & 1 & 0 \\ 0 & 0 & 1 & 0 & 0 & 1 \end{array}\right] = \left[\begin{array}{c|c} e^1 23 & e^2 23 \\ \hline I_3 & I_3 \end{array}\right]$$

and therefore, in general for an m origin, n destination T. P we may write

$$A = \begin{bmatrix} e^1 mn & e^2 mn \dots & e^m mn \\ In & In \quad \dots & In \end{bmatrix}$$

where e'mn is an m × n matrix having a row of write elements as its jth row and o's every where else and in is the n × n Identity matrix.

If aij denotes the column vector at A associated with any variable x_{ij} then it easily verified that $a_{ij} = e_j + e_{mj} - e_{ij}$ where $e_j, e_{m+j} \in R^{m+n}$ are unit vectors.

Subtracting the two, we have

$$\sum_{i=1}^{m}\sum_{j=1}^{n} xij - \sum_{j=1}^{n-1}\sum_{i=1}^{m} xij = \sum_{j=1}^{m} ai - \sum_{j=1}^{n-1} bj$$

or $$\sum_{i=1}^{m}\left(\sum_{j=1}^{n} xij - \sum_{j=1}^{n-1} xij\right) = \sum_{i=1}^{n} bi - \sum_{j=1}^{n-1} bi \quad (\in a_i = \in b) \; (\because \sum_j ai = \sum_j b_)$$

or $$\sum_{i=1}^{m} xin = bn$$

which happens to be the last destination constraint of the problem.

This indicates that it the first m + n – 1 constraints are satisfied then $\Sigma a_i = \Sigma b_j$ ensures that the (m + n)th constraint will be automatically satisfied.

This, out m + n equation, we have only (m + n – 1) linearly independent equations. Therefore, a basic feasible solution will consist of at most (m + n – 1) positive variables the rest being zero. Further a feasible solution involving exactly (m + n – 1) positive variables is known as non degenerate basic feasible solution otherwise it is said to be degenerate basic feasible.

MATRIX FORM OF T.P.

Consider the T.P. discussed in the last section. The set of constraints $\sum_{j=1}^{n} xij$ = ai and $\sum_{j=1}^{m} xij = b_j$ represent m + n equation in mn non negative variable x_{ij} each variable x_{ij} appears in exactly tow constraints one associated with the ith origin O_i and other with the jth destination D_j. In the above ordering of constraints – the origin equation first, then the destination equation – the T. P. can be restated in the matrix from as.

Minimize $z = c^T \times c' \ x \in R^{mn}$ subject to the constraints

$Ax = b, \ x \geq 0 \ b \in R^{m+n}$

Corollary (Existence of an Optimal Solution)

There always exists an optimal solution to a balanced transportation problem.

Proof:

Let $\sum_j a_i = \sum_j b$ so that a feasible solution x_{ij} exists. Therefore follows from the constraints of the problem that each x_{ij} is bounded

viz $\quad 0 \leq x_{ij} \leq \min (a, b)$

Thus the feasible region of the problem is closed, bounded and non empty and hence there exist an optimal solution.

Note:

In future we shall resume that the above condition holds for the T.P. we are discussing without mentioning it.

BASIC FEASIBLE SOLUTION OF A TRANSPORTATION PROBLEM

Since a T.P. is a spacial case of L.P.P. a basic feasible solution a transportation problem has the same definition as in for an L.P.P. However, we observe that in the case of a T.P. there are only m + n – 1 basic variable out of mn unknown, due to reducing in the constraints of the T.P.

For consider the first m + n – 1 constraints of the transportation problem:

$$\sum_{i=1}^{m} xij = bj \text{ and } \sum_{j=1}^{n} xij = a_i \ (j = 1, 2, \ldots n-1, \ i = 1, 2, \ldots m)$$

Taking summation on both sides, these yield

$$\sum_{j=1}^{n-1}\sum_{j=1}^{m} xij = \sum_{j=1}^{n-1} bj \text{ and } \sum_{i=1}^{m}\sum_{j=1}^{n} xij = \sum_{i=1}^{m} aj$$

Theorem:

(Existence of feasible solution): A necessary and sufficient condition for the existence of a feasible solution to the transportation problem (1) is that

$$\sum_{i=1}^{m} ai = \sum_{j=1}^{n} bj$$

Proof:

The condition is necessary: Let there exist a feasible solution to the T.P.

$$\text{Minimize } Z = \sum_{i=1}^{m}\sum_{j=1}^{n} xij\ cij$$

subject to the condition

$$\sum_{j=1}^{n} xij \quad a_i \qquad i = 1, 2, 3 \ldots m$$

$$\sum_{i=1}^{m} xij = b_j \quad j = 1, 2, 3 \ldots n$$

$x_{ij} \geq 0$ for all i and j

Then we have

$$\sum_{l=1}^{m}\sum_{j=1}^{n} xij = \sum_{j=1}^{m} ai \text{ and } \sum_{j=1}^{n}\sum_{i=1}^{m} xij = \sum_{j=1}^{n} bj$$

$$\text{yielding } \sum_{i=1}^{m} ai = \sum_{j=1}^{n} bj$$

The condition is sufficient: Let us assume that

$$\sum_{i=1}^{m} ai = \sum_{j=1}^{n} bj = \lambda \text{ (say)}$$

we assert that there exist a feasible solution given by $x_{ij} = a_i b_j/\lambda$ for all i and j. Clearly $x_{ij} \geq 0$ since $a_i > 0$, $b_j > 0$ for all i and j.

$$\text{Also } \sum_{j=1}^{n} xij = \sum_{j=1}^{n} (ai\, bj/\lambda) = \frac{ai}{\lambda}\cdot\left(\sum_{j=1}^{n} bj\right) = ai,\ i = 1, 2 \ldots m \text{ and}$$

$\sum_{i=1}^{m} xij = \sum_{i=1}^{m}(ai\,bj/\lambda) = b_j/\lambda\left(\sum_{j=1}^{m} ai\right) = b_j, = j = 1, 2 \ldots n$ thus x_{ij} satisfies all the constraints of the T.P hence is a feasible solution.

MATHEMATICAL FORMULATION OF ASSIGNMENT PROBLEM

Mathematically an assignment problem can be stated as follows:

Minimize the total cost

$$Z = \sum_{i=1}^{n}\sum_{j=1}^{n} c_{ij}x_{ij}$$

where x_{ij} = {1, if ith person is assigned to the jth job 0, if not subject to the conditions

(i) $\sum_{j=1}^{n} x_{ij} = 1$, (only one job is done by the person, i = 1, 2, ... ,n)

(ii) $\sum_{i=1}^{n} x_{ij} = 1$, (only one person should be assigned to the jth job, j = 1, 2, ... ,n)

Remark:

Since x_{ij} = 0 or 1 for all i and j exactly one variable in each row or column constant |s| (and other variables is zero) the structure of the problem is an nontransportation array and such three are 2n – 1 basic variables of these exactly n has the value 1 and the remaining n – 1 the value 0. Thus, the assignment problem is heavily degenerate and it will be frustrating to attempt to solve it by transportation method.

FUNDAMENTAL THEOREMS

Now we shall prove two important theorems on which the solution to an assignment problem is fundamentally based.

Theorem:

(Reduction Theorem). *In an assignment problem if we add (or subtract) a constant to every element of a row (or column) of the cost matrix [c_{ij}], then an assignment which minimizes the total cost for one matrix also minimizes the total cost for the other matrix.*

Or

Mathematical statement of reduction theorem

If $x_{ij} = X_{ij}$ *minimizes* $Z = \sum_{i=1}^{n}\sum_{j=1}^{n} c_{ij}x_{ij}$ *over all* $x_{ij} = 0$ *or* 1 *such that*

$\sum_{i=1}^{n} x_{ij} = 1$, $\sum_{j=1}^{n} x_{ij} = 1$, *then* $x_{ij} = X_{ij}$ *also minimizes*

$Z' = \sum_{i=1}^{n}\sum_{j=1}^{n} c_{ij}'x_{ij}$ *where* $c_{ij}' = c_{ij} \pm a_i \pm b_j$; b_i ; a_i, b_j *are some real numbers for* $i, j = 1, 2, \ldots, n$.

Proof:

$$\text{We have } Z' = \sum_{i=1}^{n}\sum_{j=1}^{n} c_{ij}'x_{ij}$$

$$= \sum_{i=1}^{n}\sum_{j=1}^{n} (c_{ij} \pm a_i \pm b_j)x_{ij}$$

$$= \sum_{i=1}^{n}\sum_{j=1}^{n} c_{ij}x_{ij} \pm \sum_{i=1}^{n}\sum_{j=1}^{n} a_i x_{ij} \pm \sum_{i=1}^{n}\sum_{j=1}^{n} b_j x_{ij}$$

$$= Z \pm \sum_{i=1}^{n} a_i \sum_{j=1}^{n} x_{ij} \pm \sum_{j=1}^{n} b_j \sum_{i=1}^{n} x_{ij} \qquad \left[\because Z = \sum_{i=1}^{n}\sum_{j=1}^{n} c_{ij}x_{ij}\right]$$

$$= Z \pm \sum_{i=1}^{n} a_i . 1 \pm \sum_{j=1}^{n} b_j . 1 \qquad \left[\because \sum_{i=1}^{n} x_{ij} = 1 = \sum_{j=1}^{n} x_{ij}\right]$$

$$= Z \pm \sum_{i=1}^{n} a_i \pm \sum_{j=1}^{n} b_j .$$

Since terms $\sum_{i=1}^{n} a_i, \sum_{j=1}^{n} b_j$ are independent of x_{ij}'s, if follows that Z' is minimized whenever Z is minimized and conversely.

Hence $x_{ij} = X_{ij}$ which minimizes Z well also minimize Z'.

Theorem 2.

If all $c_{ij} \geq 0$ *and there exists a solution* $x_{ij} = X_{ij}$ *which satisfies*

$\sum_{i=1}^{n}\sum_{j=1}^{n} c_{ij}x_{ij} = 0$,

then this solution is an optimal solution for the problem (i.e., minimizes the objective function).

Proof:

Since all $c_{ij} \geq 0$ and all $x_{ij} \geq 0$, the objective function $Z = \sum_{i=1}^{n}\sum_{j=1}^{n} c_{ij}x_{ij}$ cannot be negative. The minimum possible value that Z can attain is 0.

Hence the solution $x_{ij} = X_{ij}$ for which $\sum_{i=1}^{n}\sum_{j=1}^{n} c_{ij}x_{ij} = 0$ is an optimal solution.

THE ASSIGNMENT PROBLEM

The assignment problem is a special case of transportation problem in which the objective is to assign a number of assigns to the equal number of destinations at a minimum cost. The assignment is to be made on one-to-one basis.

THE NATURE OF ASSIGNMENT PROBLEMS

Let there be n jobs to be performed and for doing these jobs n persons are available.

Assume that each person can do each job at a time, though with varying degree of efficiency.

Let c_{ij} be the cost (payment) of assigning ith person to the jth job. Then the problem is to find an assignment (which job should be assigned to which person) of that the total cost for performing all jobs is minimum.

The above assignment problem can be stated in the form of n × n matrix $[c_{ij}]$ of real numbers called cost *matrix* or effectiveness *matrix* as follows:

Cost Matrix

Jobs

	1	2	3	...	j	...	n
1	c_{11}	c_{12}	c_{13}	...	c_{1j}	...	c_{1n}
2	c_{21}	c_{22}	c_{23}	...	c_{2j}	...	c_{2n}
3	c_{31}	c_{32}	c_{33}	...	c_{3j}	...	c_{3n}
⋮	⋮	⋮	⋮		⋮		⋮

	⋮	⋮	⋮	⋮		⋮		⋮
Persons	i	c_{i1}	c_{i2}	c_{i3}	...	c_{ij}	...	c_{in}
	⋮	⋮	⋮	⋮		⋮		⋮
	n	c_{n1}	c_{n2}	c_{n3}	...	c_{nj}	...	c_{nn}

ASSIGNMENT ALGORITHM (HUNGARIAN ASSIGNMENT METHOD)

Various steps of the computational procedure for obtaining an optimal assignment are as follows.

Step 1. Modify cost matrix by subtracting the minimum element of each row of the cost matrix, from all the elements of respective rows. Further, modify the resulting matrix by subtracting the minimum element of each column from all the elements of the respective columns. These operations create zeros.

Step 2. In the modified matrix search for an optimal assignment.

Starting with row 1 or the matrix obtained in step 1, examine rows successively until a row with exactly one zero element is found. Mark '□' at this zero as an assignment will be made there. Mark '×' at all other zeros if lying in the column containing the assigned zero. This eliminates the possibility of marking further assignments in that column. Continue in this manner until all the rows have been examined.

When the set of rows has been completely examined, an identical procedure is applied successively to columns. Starting with column 1, examine all the columns until a column containing exactly one unmarked zero is found. Then make an assignment in that position (indicated by □) and mark '×' at all zeros in the row containing this marked zero. Proceed in this way until the last column is examined.

Continue the above operations on rows and columns successively until we reach to any of the two situations

(i) all the zeros have been marked '□' or '×'

(ii) the remaining zeros lie at least two in each row and column.

In situation (i), we have a maximal assignment (assignment as much s we can) and in situation (ii) still we have some zeros to be treated. To work with such situations of zeros there is again an algorithm, complicated enough. But to avoid this highly complicated algorithm we use the trial and error method to break up such ties of zeros.

Now there are tow possibilities:

(i) If there is an assignment in every row and every column (i.e., total number of marked '□' zeros is exactly n), then we have obtained a complete optimal assignment plan.

(ii) If every row and every column do not contain an assignment (i.e., total number of marked '□' zeros is less than n), then we shall modify the cost matrix by creating some more zeros in it.

Step 3. If in step 2 every row and every column of the matrix do not contain assignment then draw the minimum number of horizontal and vertical lines to cover all the zeros at least once in the resulting matrix.

Rule to Draw Minimum Number of Lines

(i) Mark (√) all rows that do not have any marked '□' zero.

(ii) Mark (√) columns which have zeros in marked rows.

(iii) Mark (√) rows (not already marked) which have assignments in marked columns.

(iv) Repeat steps (ii) and (iii) until the chain of marking ends.

(v) Draw lines through unmarked rows and through marked columns.

This will give us the minimal system of lines.

Notes:

1. The lines thus drawn (horizontal and vertical both) are the minimum number of lines to pass through all the zeros of the matrix. It can be shown that the minimum number of lines required to pass through all the zeros of the matrix is the same as the maximum number of assigned independent zeros of the matrix.

   Thus, if the number of lines is exactly n, then the complete assignment plan is obtained wile if the number of lines is less than n, then the complete assignment is not possible.

2. These lines cover all the zeros and each line passes through one and only one marked zero (assignment). If there are two marked '□' zeros in a row then it follow that we are assigning two jobs to one

person which is a violation of the hypothesis. Thus no line passes through more than one marked '□' zero.

Step 4. Select the smallest of the elements that do not have a line though them, subtract if from all the elements that do not have a line through them, add it to every element that lies at the intersection two lines and leave the remaining elements of the matrix unchanged. In the modified matrix number of zeros are increased (never decrease) than that in step 2. Now apply the step to this new matrix. If still a complete optimal assignment is not possible in this matrix, then repeat steps 3 and 4 iteratively. Continue the process until minimum number of lines be n.

Thus exactly one marked '□' zero in each row and each column of the matrix is obtained. The assignment corresponding to these marked '□' zero in each column of the matrix is obtained. The assignment corresponding to these marked '□' zeros will give the optimal assignment.

Note: The procedure of subtracting the minimum element of all uncovered elements from all such elements and adding this minimum element to the elements placed at the intersection does not change the optimum solution i.e., the two matrices will have the same optimal solution. For the above mentioned tow operations of addition and subtraction are the resultant of operations of subtracting the above chosen minimum element from the uncrossed rows and adding it to all the crossed columns and such operations do not change the optimum solution.

TIME MINIMIZATION PROBLEM

In the Time Minimizing Transportation Problem, the objective is to minimize the time of transportation rather than the cost of transportation. Such problems are often encountered in military, hospital, management, five service etc.

The minimization transportation problems are similar to the problems discussed for, except that the unit time t_{ij} required to transport the items from the origin to the destination.

The systematic procedure for the solution of such problems is as follows:

Step 1. Determine an initial basic feasible solution by using any methods.

Step 2. Determine ^{T}j for this basic feasible solution and cross out all the non basic calls for which $t_{ij} \geq {}^{T}j$

Step 3. Construct a loop for the basic cells corresponding to T_j in such a way that when the values cut the corner cells are shifted around, the value this cell tends towards (not necessarily zero) zero and no variable becomes zero. It no such closed path can be formed, the solution under test is optimal, otherwise move to next step.

Step 4. Go to step 2 and repeat the procedure until an optimum basic feasible solution is attained.

TRANSSHIP MINT PROBLEM

A transportation problem in which available commodity may not be sent directly from which available commodity may not be sent directly from sources to destination is termed as a transshipment problem. Such

2						2
	7		3		4	
7		1		6		
	7		1		3	5t ∈ = 5
				5		
	3		4		6	
	4	1			5t ∈ = ∈	

Notice that even after the inclusion of cell (2, 3) in the basis, the basic cells do not form a loop. The net evaluation can now be computed and solution is tested for optimality.

STARTING TABLE

Since all the net evaluations for the non basic variables are not non positive, the initial solution is not optimum. The non basic call (1, 3) must enter the basis: The exist criterion removes basis cell (2, 3) from the basis [Max. $\theta = \in$]

2 −θ			(3)	θ	(4)	ui
	7		3		4	5
2 +θ		1		∈ −θ		
	2		1		3	0

	(2)	(0)	5	3
	3	4	6	
v_j	2	1	2	

Starting table

First Iteration

Induced the cell (1, 3) into the basis and drop the cell (2, 3) from it Determine the current net evaluations. Since all of them are not non positive, the current solution can be improved.

2 −θ			(3) ∈ +θ			u_i
	7		3		4	0
2		1			(−4)	−5
	2		1		3	
θ	(6)		(4)	5 −θ		2
	5		4		6	
v_j	7	6		4		

First iterated table

Second Iteration

Introduced the cell (3, 1) and drop the cell (1, 1) from the basis since some of the current net evolution are still positive, the current solution can for there be improved.

	(−6)		(−3)	2		u_i
	7		(3)		4	−1
2 −θ		1		θ	(2)	
	2		1		(3)	0
2 +θ		(−2)		3 −θ		
	3		(4)		6	2
v_j	2		1	5		

Second Iteration table:

Third Iteration

Introduced the cell (2, 3) and drop the cell (2, 1) from the basis. Since all the current net evaluation are non positive the current solution is an optimum one.

	(–6)	(–1)	2		u_i
	7	(3)		4	4
	(–2) 1		2		3
	2	1		3	
4		(0)	1		
	5	(4)		6	6
v_j	–3	–2	0		

Optimum table.

UNBALANCED TRANSPORTATION PROBLEM

Some time there is more demand than the availability and *vice-versa* such type of transportation problem are called unbalanced transportation problems.

If in a transportation problem

$$\sum_{i=1}^{m} ai \neq \sum_{i=1}^{n} bj$$

Then the problem is called unbalances transportation problem. There are two type of unbalance problem.

Case I: (Excess Availability) : When never $\Sigma a_i - \Sigma b_j$ we introduced a dummy destination in a transportation table the cost of transporting to this destination are all set equal to zero. The requirement at this dummy destination is assumed to be equal to $\Sigma a_i - \Sigma b_j$

Example 1:

Solve the transportation table.

	D_1	D_2	D_3	*available*
	20	*30*	*10*	*800*
	5	*15*	*25*	*500*
Required	*300*	*300*	*400*	

Solution:

Here the total available exceed the total demand. We consider a fictitious destination requiring $\sum_{i=1}^{m} a_i - \sum_{j=1}^{n} b_j = 1300 - 1000 = 300$

units of quantity. The new transportation problem is

	D_1	D_2	D_3	fictitious	available
O_1	30	20	10	0	800
O_2	5	15	25	0	500
Required	300	300	400	300	1300

Solve this by usual method; we get are optimal solution given by the following table

	D_1	D_2	D_3	fictitious	available
O_1		100	400	300	800
O_2	300	200			500
Required	300	300	400	300	

The table indicate that the 300 units will be less at O_1 not dispatched.

Case II: (Shortage in Availability) : Whenever $\Sigma a_i \leq \Sigma b_j$, introduced a dummy source in the transportation table. The cost of transporting from this source to any destination any all set equal to zero the availability at this dummy source is assumed to be equal to

$$\Sigma b_j - \Sigma a_i$$

to or (subtracted from) a non zero quantity, we need to write Σ only when it appears alone.

Remark:

A degenerate B.F.S. can also be converted into a non degenerate B.F.S. by introducing a small positive quantity say I in one or more empty (independent) cells and correspondingly increasing the capacity and demand by the same quantity say in one or more empty (independent) cells and correspondingly increasing the capacity and demand by the same quantity 1.

Example 2:

Obtain an optimum basic feasible solution to the following degenerate T.P.

To

	7	3	4	2
from	2	1	3	3 *available*
	3	4	6	5
Demand.	4	1	5	10

Solution:

The initial solution by north west carnivore is

7	7			5		4	2
2		1					
2		1			3	3	
					5		
	3			4		6	5
	4		1		5		

In order to complete the basic and there by remove degeneracy we rigger only one more positive basic variable we select the variable x_{23} and allocate a negligibly small positive quantity in the call (2, 3) as shown in the table below.

Degeneracy in Transportation Problem

A feasible solution to a transportation problem is degenerate if less than m + n – 1 of the x_{ij} are positive. Degeneracy may be encountered in the process of determining the initial basic feasible solution or at some subsequent iteration. The reason for degeneracy is that the sum of certain number of a_i is exactly equal to the sum of some b_j is . Due to this property, while finding a F.S. the allocation in some cell, satisfies the column as well as row requirements simultaneously.

In such case we chose one or more as many individual allocations are required to have exactly m + n – 1 allocations) empty cell of the matrix (generally lowest cost cells is suits) in a way that if these calls are allocated, they form an independent set of allocations. In these chosen cell or cells, we may assign one or very small positive quantity '∈' (epsilon), governed by the following rules:

(i) $O < \in < x_{ij}$

(ii) $O + \in = \in$

(iii) $x_{ij} \pm \in = x_{ij}$

Further, as long as it is needed, this $\in$ is to be considered as a real allocation through all the subsequent iteration. Above rule show that even after introducing $\in$, the original problem is not changed. It is merely a technique to apply the optimality test. As $\in$ has no physical significant ultimately it is to be omitted.

In numerical problem due to the property

(3) Of $\in$, we need not to write it whenever it is added.

THE TRANSPORTATION ALGORITHM

The Algorithm

Various steps are as follows :

Step 1. Construct a transportation table, entering the origin capacities a_i, the destination requirements b_j and the cost c_{ij}.

Step 2. Find the initial basic feasible solution. Enter the solution in the upper left corners of the basic cells.

Step 3. Find the set of numbers u_i and ui for each row and column s.t. for all the occupied cells $C_{ij} = u_i + v_j \; \forall_{i,j}$ for which (i, j) is in the basis starting initially with some $u_i = 0$ and entering successively the value of u_i and v_j on the transportation table.

Step 4. Computer the net evaluation $Z_{ij} - c_{ij} = u_i + v_j - c_{ij}$ for all the non basic cells and enter them in the upper right corners of the corresponding cells.

Step 5. Examine the sign of each $Z_{ij} - c_{ij}$. It all $Z_{ij} - c_{ij} \leq 0$, then the current boric feasible solution is an optimum one. It at least one $Z_{ij} - c_{ij} > 0$, select the variable x_{rs} having the largest positive net evaluation to enter the basis.

Step 6. Let the variable x_{rs} enter the basis. Allocate an unknown quantity say θ, to the cell (r, s). Identity a loop that an starts and end at the cell (r, s) and connect some of the basic cells. Add and subtract interchangeably, θ to and from the transition cells of the loop in such a way that the rim requirements remain satisfied.

Step 7. Assign a maximum value θ in such a way that the value of one basic variable be comes zero and the other bane variables remain non-negative. The basic cell whose allocation has been reduced to zero, leaves the basis.

Step 8. Return to step 3 and repeat the process until an optimum basic feasible solution has been obtained.

The Column Minima Method

Step 1. Determine the smallest cost in the first column of the transportation table. Let it be c_{ij}. Allocate x_{ij} = min (a_1, b_1,) in the call (i, 1)

Step 2. If $x_{ij} = b_1$, cross off the first column of the transportation table and move towards right to the second column.

If $yi_1 = ai_1$ cross off the ith row of the transportation table and reconsider the first column with the remaining demand.

If $xj_1 = b_1 = ai$, cross off the ith row and make the second allocation $yk_1 = 0$ in the cell (k, 1) with ck_1 being the new minimum cost in the first column. Cross of the column and move towards right to the second column.

Step 3. Repeat step 1 and 2 for the resulting reduced transportation table until all the rim requirements are satisfied.

The Row Minima Method

Step 1. The smallest cost in the first row of the transportation table determined. Let it be C_{ij}. Allocate the maximum decibel amount x_{ij} = min (a, bj) is the cell (i, j) so that either the capacity of origin O_1 is exhausted, or the requirement at destination D, is satisfied or both.

Step 2. If $x_{1j} = a_1$ so that the availability at origin O_1 is completely exhausted, cross off the first row of the table and move down to the second now.

If $x_{ij} = b_j$ so that the requirement at destination as satisfied, cross off the jth column and reconsider the first row with the remaining availability of origin O_1.

If $x_{ij} = a_i = b_j$ the origin capacity O_1 is completely exhausted as well as the requirement at destination D_j is completely satisfied an arbitrary tie breaking choice is made. Cross off the jth column and made the second allocation $x_{ik} = 0$ in the cell (1, k) with c_1k being the new minimum cost in the first row. Cross off the first row and move down to the second row.

Step 3. Repeat step 1 and 2 for the resulting reduced transportation table until all the rim requirements are satisfied.

TRIANGULAR BASIS IN T.P.

The reader may recall that a system of n linear equations Ax = b is called a triangular system it the matrix A is triangular i.e., it is the form.

$$A = \begin{pmatrix} a_{11} & a_{12} & a_{13} & a_{1n} \\ 0 & a_{22} & a_{23} & a_{2n} \\ 0 & 0 & a_{33} & a_{3n} \\ 0 & 0 & 0 & a_{nn} \end{pmatrix} a_{ij} \neq 0 \text{ for each i}$$

The system has a property that there is at least one equation having only one variable. This variable can be eliminated from the remaining equations resulting in a reduced system of (n – 1) equations which again happens to be triangular. Thus a triangular system can be easily solved by back substitution.

Definition: (Triangular Basis) : A basis for the system Ax = b is said to be a triangular basis if a Triangular system is obtained when all the non basic variables are set zero in the system.

Example 1:

A car hire company has one car at each of five depots a, b, c, d and e. A customer requires a car in each town namely A, B, C, D and E. Distance (in kms) between depots (origins) and towns (destinations) are given in the following distance matrix:

	a	*b*	*c*	*d*	*d*
A	*160*	*130*	*175*	*190*	*200*
B	*135*	*120*	*130*	*160*	*175*
C	*140*	*110*	*155*	*170*	*185*
D	*50*	*50*	*80*	*80*	*110*
E	*55*	*35*	*70*	*80*	*105*

How should cars be assigned to customers so as to minimize the distance travelled?

Solution:

Step 1. Subtracting the minimum element of each row form every element of the corresponding row, the matrix reduces to

30	0	45	60	70
15	0	10	40	55
30	0	45	60	75
0	0	30	30	60
20	0	35	45	70

Now subtracting the minimum element of each column from every element of the corresponding column, the matrix reduces to

30	0	35	30	15
15	0	0	10	0
30	0	35	30	20
0	0	20	0	5
20	0	25	15	15

Step 2. Now we give the zero assignments in our usual manner,. Row 1 has a single zero in column 2. Make an assignment by marking '□' around it and delete other zeros (if any) in column 2 by marking '×', Examining the set rows completely, an identical procedure is applied successively to columns.

30	0	35	30	15
15	⊗	0	10	⊗
30	⊗	35	30	20
0	⊗	20	⊗	5
20	⊗	25	15	15

Now column 1 has a single zero in row 4. Make an assignment by marking '□' at this zero an cross the other zero of row 4 which in not yet crossed. Column 3 has a single zero in row 2, make an assignment at this zero by putting '□' and cross the other zero of row which is not yet crossed. At this stage all zeros have been either assigned or crossed out. It is observed that row 3, 5, column 4 and column 5 each has no assignment. Hence the required solution cannot be obtained at this stage. So we proceed to the next step.

Step 3. In this step we draw minimum number of lines to cover all zeros a least once. For this we proceed as follows:

(i) Mark (√) row 3 and row 5 as they have no assignments.

(ii) Mark (√) column 2 as having zeros in the marked rows 3 and 5.

(iii) Mark (√) row 1 as it contains assignment in the marked column 2.

Now further rows or colons will be required to mark during this procedure.

(iv) Now draw line L_1 through marked column 2. Then draw lines L_2 and L_3 through unmarked rows 2 and 4.

The required lines will be L_1, L_2 and L_3. No zero is left uncovered.

		L_1				
	30	0	35	30	15	√ 4
L_2	15	⊗	0	10	⊗	
	30	⊗	35	30	20	√ 1
L_3	0	⊗	20	⊗	5	
	20	⊗	25	15	15	√ 2
		√ 3				

Step 4. In this step we select the smallest element among all uncovered elements of the matrix of step 3.

Here this element is 15. Subtracting this element 15 from all the elements that do not have a line through them and adding to every element that lies at the intersecting of two lines and leaving the remaining elements unchanged we get the following matrix.

15	0	20	15	0
15	15	0	10	0
15	0	20	15	5

0	15	20	0	5
5	0	10	0	0

Step 5. Now again performing the step 2 we make the zero assignments. It is observed that there are no remaining zeros and every row (column) has an assignment as shown in the table.

15	⊗	20	15	0
15	15	0	10	⊗
15	0	20	15	5
0	15	20	⊗	5
5	⊗	10	0	⊗

Thus the complete optimal assignment plan is given by

A → e, B → c, C → b, D → a, E → d.

Form the original matrix, the minimum cost (distance travelled)

= (200 + 130 + 110 + 50 + 80) kms. = 570 kms.

Example 2:

A department head has four subordinates, and four tasks have to be performed. Subordinates differ in efficiency and tasks differ in their intrinsic difficulty. Time each man would take to perform each task is given in the electiveness matrix below. How should the tasks be allocated, one to a man, so as to minimize the total man hours?

		Subordinates			
		I	*II*	*III*	*IV*
	A	*8*	*26*	*17*	*11*
Tasks	*B*	*13*	*28*	*4*	*26*
	C	*38*	*19*	*18*	*15*
	D	*19*	*26*	*24*	*10*

Solution:

We shall solve this problem step by step to understand the method described above.

Step 1. Subtracting the minimum element of each row from every element of the corresponding row, the matrix reduces to

	I	II	III	IV
A	0	18	9	3
B	9	24	0	22
C	23	4	3	0
D	9	16	14	0

Now subtracting the minimum element of each column form every element of the corresponding column, the matrix reduces to

	I	II	III	IV
A	0	14	9	3
B	9	20	0	22
C	23	0	3	0
D	9	12	14	0

Step 2. Now test whether it is possible to make an assignment using only zeros.

Starting with row 1 of the matrix examine the rows one by one until a row containing exactly single zero element if found. We mark '□' at this zero, i.e., make an assignment and mark a cross '×' over all zeros if lying in the column continuing the assigned zero. Continue in this manner until all the rows have been examined. The illustration of the procedure is shown in he following table.

	I	II	III	IV
A	0	14	9	3
B	9	20	0	22
C	23	0	3	⊗
D	9	12	14	0

Now starting with column 1, examine all the columns until a column containing exactly one zero is found. We mark '□' in that position i.e., make an assignment cross other zeros if lying in the row containing this marked zero.

Continue in this manner until all the columns have been examined. The illustration of the procedure is shown in the following table.

	I	II	III	IV
A	0	14	9	3
B	9	20	0	22
C	23	0	3	⊗
D	9	12	14	0

At this stage all zeros have been either assigned or crossed out. We observe that every row and every column have one assignment, so we have the complete 'zero assignment' given by

A → I, B → III, C → II, D → IV.

This assignment is also optimal for the original matrix.

The minimum total man hours = 8 + 4 + 19 + 10 = 41.

Example 3:

Solve assignment problem represented by the following matrix:

	I	*II*	*III*	*IV*	*V*	*VI*
A	*9*	*22*	*58*	*11*	*19*	*27*
B	*43*	*78*	*72*	*50*	*63*	*48*
C	*41*	*28*	*91*	*37*	*45*	*33*
D	*74*	*42*	*27*	*49*	*39*	*32*
E	*36*	*11*	*57*	*22*	*25*	*18*
F	*3*	*56*	*53*	*31*	*17*	*28*

Solution:

Step 1. Subtracting the minimum element of each row from every element of the corresponding row and then subtracting the minimum element of each column form every element of the corresponding column, the matrix reduces to

0	13	49	0	0	13
0	35	29	5	10	0
13	0	63	7	7	0
47	15	0	20	2	0
25	0	46	9	4	2
0	53	50	26	4	20

Step 2. Make the 'zero assignments' in usual manner. The illustration is shown in the following table. Since row 3 and column 5 have no assignments so we proceed to the next step.

$\otimes$	13	49	0	$\otimes$	13
$\otimes$	35	29	5	10	0
13	$\otimes$	63	7	$\otimes$	$\otimes$
47	15	0	20	2	$\otimes$
25	0	46	9	4	2
0	53	50	26	4	20

Step 3. Draw minimum number of lines to cover all zeros at least once. For this we proceed as follows:

(i) Mark ($\sqrt{}$) row 3 as having no assignment.

(ii) Mark ($\sqrt{}$) columns 2 and 6 as having zeros in marked row 3.

(iii) Mark ($\sqrt{}$) rows 5 and 2 as having assignments in the marked columns 2 and 6.

(iv) Mark ($\sqrt{}$) column 1 (not already marked) as having zero in the marked row 2.

(v) Then mark ($\sqrt{}$) row 6 as having assignment in the marked column 1.

	L_1	L_2				L_3	
L_4	$\otimes$	13	49	0	$\otimes$	13	
	$\otimes$	35	29	5	10	0	$\sqrt{}$ 5
	13	$\otimes$	63	7	7	$\otimes$	$\sqrt{}$ 1
L_5	47	15	0	20	2	$\otimes$	
	25	0	46	9	4	2	$\sqrt{}$ 4
	0	53	50	26	4	20	$\sqrt{}$ 7
	$\sqrt{}$	$\sqrt{}$				$\sqrt{}$	
	6	2				3	

Now draw lines L_1, L_2, L_3 through marked columns 1, 2, 6 respectively and L_4, L_5 through unmarked rows 1, 4 respectively.

This way minimum set of five lines (5 < 6) to cover all the zeros is obtained.

Step 4. Now the smallest element among all uncovered elements is 4. Subtracting this element 4 from all the uncovered elements, adding to every element that lies at the intersection of two lines and leaving the remaining elements unchanged the matrix of step 3 reduces to the new form as shown in the table.

4	17	49	0	0	17
0	35	25	1	6	0
13	0	59	3	3	0
51	19	0	20	2	4
25	0	42	5	0	2
0	53	46	22	0	20

Step 5. Repeating the step 2, make the 'zero assignments' as shown in the following table.

Thus exactly one marked '□' zero in each row and each column of the matrix is obtained.

4	17	49	0	⊗	17
0	35	25	1	6	⊗
13	⊗	59	3	3	0
51	19	0	20	2	4
25	0	42	5	⊗	2
⊗	53	46	22	0	20

Thus the optimal assignment is

A → IV, B → I, C → VI, D → III, E → II, F → V.

From the original matrix, minimum cost = 11 + 43 + 33 + 27 + 11 + 17 + 142.

Note:

Another optimal solution this assignment problem is shown in the following table i.e.,

A → IV, B → VI, C → II, D → III, E → V, F → I.

4	17	49	0	⊗	17
⊗	35	25	1	6	0

13	0	59	3	3	⊗
51	19	0	20	2	4
25	⊗	42	5	0	2
0	53	46	22	⊗	20

From the original matrix, minimum cost = 11 + 48 + 28 + 27 + 25 + 3 = 142.

Example 4:

Solve the minimal assignment problem whose effectiveness matrix is

	I	*II*	*III*	*IV*
A	*2*	*3*	*4*	*5*
B	*4*	*5*	*6*	*7*
C	*7*	*8*	*9*	*8*
D	*3*	*5*	*8*	*4*

Solution:

Step 1. Subtracting the minimum element of each row from every element of the corresponding row, the matrix reduces to

0	1	2	3
0	1	2	3
0	1	2	1
0	2	5	1

Now subtracting the minimum element of each column from every element of the corresponding column, the matrix reduces to

0	0	0	2
0	0	0	2
0	0	0	0
0	1	3	0

Step 2. Now test whether it is possible to make an assignment using only zeros. Here none of the rows or columns contain exactly one zeros, Therefore, we start with row 1 searching two zeros. While examining rows successively, it is observed that row 4 has two zeros. Now, arbitrarily make an assignment (indicated

by □) to one of these two zeros, say zeros say zero in the column 1 and cross other zeros in row 4 and column 1. Now we find column 1 which contains only one unmarked zero in row 3. We make assignment (indicated by □) at this zero and cross all other zeros of this row. Now again we check the rows and columns for one unmarked zero. There is no such row or column. So we start with row 1 searching two unmarked zeros, say zero of column 2, and cross other zeros of row 1 (not already crossed) and column 2. Now the second row contains only one unmarked zero in third column where we can make an assignment (indicated by □).

At this stage all zeros have been either assigned or crossed out. We observe that every row and every column have one assignment, so we have the complete 'zero assignment'.

Tables 1, 2, 3 show the necessary steps for reaching the optimal assignment.

Table 1

⊗	0	0	2
⊗	0	0	2
⊗	0	0	0
0	1	3	⊗

Table 2

⊗	0	0	2
⊗	0	0	⊗
⊗	⊗	⊗	0
0	1	3	⊗

Table 3

⊗	0	⊗	2
⊗	⊗	0	2
⊗	⊗	⊗	0
0	1	3	⊗

Thus we get the following optimal assignment ;

$$A \to II,\ B \to III,\ C \to IV,\ D \to I.$$

Minimum cost = 3 + 6 = 8 = 3 = 20.

Note. In this example other optimal assignments are also possible. Students must try to find them. Each has the cost 20.

6

Simplex Method with Optimal Solutions

INTRODUCTION

Fundamental theorem of linear programming makes it easy to find an optimal solution because it deals with basic feasible solutions only. But it is also not easy job to enumerate all the B.F.S. even for small values of m (number of constraints) and n (number of variables). To overcome this difficulty Simplex method or Simplex Algorithm was developed by George Dentist in 1947. This method provides an efficient technique which can be applied for solving linear programming problems of any magnitude (involving two or more decision variables). The simplex method is an iterative procedure, for finding, in a systematic manner the optimal solution to a linear programming problem. The simplex algorithm according to its iterative search selects the optimal solution from among the set of feasible solutions to the problem.

CONDITION FOR IMPROVED BASIC FEASIBLE SOLUTION TO BECOME OPTIMAL

Theorem:

Let $x_B = B^{-1} b$ be the basic (degenerate or non-degenerate) feasible solution of the L.P.P. $Z = c\,x$ s.t. $A\,x = b$, $x \geq 0$. Let $Z^* = c_B\,x_B$ be the value of the objective function at any iteration of simplex method. If $c_j - Z_j \leq 0$ for every column α_j in A but not in B then Z^* is the optimum value of the objective function Z and x_B is an optimal basic feasible solution.

Proof:

Suppose $x_B = (x_{B1}, x_{B2}, \ldots, x_{Bm})$ is the B.F.S. of the given L.P.P. We have

$A = (\alpha_1, \alpha_2, \ldots, \alpha_N)$, $N = m + n$

and the basis matrix $B = (\beta_1, \beta_2, \ldots, \beta_m)$.

$\therefore B_{mB} = b$ or $x_B = B^{-1} b$. ...(1)

The value of the objective function for the B.F.S. xB is

$$Z^* = c_B\, x_B = \sum_{i=1}^{m} c_{Bi} x_{Bi}. \qquad \text{...(2)}$$

Also, we are given that $c_j - Z_j \leq 0$ for every column α_j in A but not in B.

Let 'x' = $[x_1', x_2', \ldots, x_N']$ be any feasible solution and Z' be the value of the objective function for this solution.

Then we have

$$A\,x' = b \qquad \text{...(3)}$$

and $$Z' = c\,x' = \sum_{j=1}^{N} c_j x_j' \qquad \text{...(4)}$$

Now Z^* will be the optimum value of the objective function if we have $Z^* \geq Z'$. To prove $Z^* \geq Z'$ we proceed in the following manner.

We have $x_B = B^{-1} b = B^{-1} (A\,x')$, from (3)

$= (B^{-1} A)\, x'$

$= Y\, x'$, where $Y = [Y_1, Y_2, \ldots, Y_N]$

$= [Y_1, Y_2, \ldots, Y_N]\, [x_1', x_2', \ldots, x_N']$

$$= \begin{bmatrix} y_{11} & y_{12} & \cdots & y_{1j} & \cdots & y_{1N} \\ y_{21} & y_{22} & \cdots & y_{2j} & \cdots & y_{2N} \\ \cdots & \cdots & \cdots & \cdots & \cdots & \cdots \\ \cdots & \cdots & \cdots & \cdots & \cdots & \cdots \\ y_{m1} & y_{m2} & \cdots & y_{mj} & \cdots & y_{mN} \end{bmatrix} \begin{Bmatrix} x_1' \\ x_2' \\ \vdots \\ x_j' \\ \vdots \\ x_N' \end{Bmatrix}$$

or $[x_{B1}, x_{B2}, \ldots, x_{Bi}, \ldots, x_{Bm}]$

$$= \left[\sum_{j=1}^{N} y_{1j} x_j', \sum_{j=1}^{N} y_{2j} x_j', \ldots, \sum_{j=1}^{N} y_{ij} x_j', \ldots, \sum_{j=1}^{N} y_{mj} x_j' \right]$$

Equating ith component on both sides, we get

$$x_{Bi} = \sum_{j=1}^{N} y_{ij} x_j'. \qquad ...(5)$$

We have $c_j - Z_j \le 0$ for all j for which α_j is not in B.

If this inequality also holds for all those j for which α_j is in B then $c_j - Z_j \le 0$ for all j for which α_j also belong to B.

Now if $\alpha_j = B_i$, then

$$\alpha_j = \beta_i = 0.\beta_1 + 0.\beta_2 + ... + 0.\beta_{i-1} + 1.\beta_i + ... + 0.\beta_m$$

which shows that $Y_j = e_i$, a vector whose ith component is unity.

Again $\quad \alpha_j = \beta_i$ gives $c_j = c_{Bi}$

$\therefore \; c_j - Z_j = c_j - c_B Y_j = c_j - c_B e_i = c_j - c_{Bi} = 0$

Thus $c_j - Z_j = 0$ for all those j for which $\alpha_j \in B$.

$\therefore c_j - Z_j \le 0$ for all α_j in A

$\Rightarrow c_j \le Z_j$

$\Rightarrow c_j x_j' \le Z_j x_j', \because x_j' \ge 0$

$$\Rightarrow \sum_{i=1}^{N} c_j x_j' \le \sum_{j=1}^{N} Z_j x_j'$$

$$\Rightarrow Z' \le \sum_{j=1}^{N} x_j'(c_B Y_j), \text{ from (4)}$$

$$= \sum_{j=1}^{N} x_j' \left(\sum_{i=1}^{m} c_{Bi} y_{ij} \right)$$

$$= \sum_{i=1}^{m} c_{Bi} \left(\sum_{j=1}^{N} x_j' y_{ij} \right)$$

$$= \sum_{i=1}^{m} c_{Bi} x_{Bi}, \text{ from (5)}$$

$\therefore Z' \le Z^*$.

Hence Z^* is the optimum value of the objective function.

ALTERNATIVE OPTIMAL SOLUTIONS

The given linear programming problem is said to possess an alternative optimal solution if the set of variables giving the optimal value of the objective function is not unique.

Theorem:

(Conditions for Alternative Optimum Solution)

Suppose there exists an optimum B.F.S. to a L.P.P. and

(i) if for some a_j in A but not in B, $c_j - Z_j = 0$ and y_{ij} £ 0 for all $i = 1, 2, ..., m$ then a non-basic alternative solution will exist.

(ii) if for some a_j in A but not in B, $c_j - Z_j = 0$ and $y_{ij} > 0$ for all least one i then an alternative basic optimum solution will exist.

Proof:

(i) We have discussed in 6 that if we introduce the column vector a_j in B where a_j is in A but not in B and y_{ij} £ 0 for all $i = 1, 2, ... ,m$ then a non-basic feasible solution x_B' with $(m + 1)$ number of positive variables is given by

$$x_B' = [x_{B1}', x_{B2}', ... ,x_{Bm}', l]$$

where $x_{Bi}' = x_{Bi} - \lambda y_{ij}$, $i = 1, 2, ... ,m$, $l > 0$

and the value of the objective function for this new F.S. is given by

$$Z' = Z + l\,(c_j - Z_j).$$

If $c_j - Z_j = 0$, then $Z' = Z$

i.e., the value of the objective function for this new non-basic F.S. is also equal to Z (optimal value). Hence this new non-basic F.S. is an alternative optimal solution of the given L.P.P.

(ii) We have shown in theorem of 5 that if $y_{ij} > 0$ for at least one $i = 1, 2, ..., m$ then by palpation one column br in B by the column aj which is in A but not in B, we obtain a new B.F.S. x_B' given by

$$x_B' = [x_{B1}', x_{B2}', ... ,x_{Bm}']$$

where $x_{Bi}' = x_{Bi} - \dfrac{y_{ij}}{y_{rj}} x_{Br}$, $i = 1, 2, ... ,m$, $i \geq r$

$$x_{Br}' = \frac{x_{Br}}{y_{rj}},\ i = r$$

and $$\frac{x_{Br}}{y_{rj}} = \underset{i}{\text{Mini}}\left(\frac{x_{Bi}}{y_{ij}}, y_{ij} > 0\right).$$

The value of objective function for this new B.F.S. is given by

$$Z' = Z + \frac{x_{Br}}{y_{rj}}(c_j - Z_j)$$

i.e., the value of the objective function for this new B.F.S. is also equal to Z' (optimal value).

Hence this new B.F.S. is an alternative optimal F.F.S.

INCONSISTENCY AND REDUNDANCY IN CONSTRAINT EQUATIONS

Redundancy is constraint equations.

By redundancy in constraint equations we mean that the system has more than enough number of constraint equations, in other words it has more constraint equations than the number of variables.

This is the situation when

$$r(A) = r(A\ b) = k \leq n < m.$$

In this case there will be (m – k) redundant equations.

Inconsistency

As already defined, the set of constraints (linear equations) is said to be inconsistent if $r(A) \neq r(Ab)$. Before solving a LP. problem by simplex method, we should have $r(A) = r(Ab)$ *i.e.,* the constraint aquariums (after introducing the slack and artificial variables) shows be consistent. Since in simplex method we always have $r(A) = r(A\ b) = m$.

If the system $A\ x = b$ involves artificial variables, then we cannot say whether this system is consistent or there is any redundancy. Below we give the cases (without proof) to decide about the consistency and redundancy in such systems.

Case I.

If the basis B contains no artificial vector and the optimality condition is satisfied (at any iteration) then the current solution is a B.F.S. of the problem.

Case II.

If one or more artificial vector appears in the basis B at a zero level i.e., the value of the artificial variables corresponding to artificial vectors *in B are zero and the optimality condition is satisfied (at any iteration) then the system is consistent. Furthermore if $y_{ij} = 0 \ \forall j$ and $x_{Br} = 0$ and r corresponds to the row containing an artificial vector, then the rth constraint equation is* **redundant**.

Case III.

If at least one artificial vector appears in the basis B at a positive level i.e., the value of at least one artificial variables corresponding to artificial vector in B is non-zero and the optimality condition is satisfied (at any iteration), then there exists **no feasible** *solution of the problem.*

TO DETERMINE STARTING B.F.S.

In this article we are going to discuss a method for finding a most convenient initial B.F.S. to a linear programming problem. In the constraints of a general linear programming problem three may be any of the three signs $\leq = \geq$. Let us assume that the requirement vector b ≥ 0 (if any of the b_i's is negative, multiply the corresponding constraint by – 1.)

Case 1.

To find initial B.F.S. when at the original constraints have $\leq$ sign

To convert all the constraints into equations we insert slack variables only .the equations obtained are as follows ..

$$a_{11}x_1 + a_{12}x_2 + \dots + a_{1n}x_n + 1.x_{n+1} = b_1$$
$$a_{21}x_1 + a_{22}x_2 + \dots + a_{2n}x_n + 1.x_{n+2} = b_2$$
$$\dots \quad \dots \quad \dots \quad \dots \quad \dots \quad \dots$$
$$\dots \quad \dots \quad \dots \quad \dots \quad \dots \quad \dots$$
$$a_{m1}x_1 + a_{m2}x_2 + \dots + a_{mn}x_n + \dots + 1.x_{n+m} = b_m$$

Here x_{n+1}, x– = 2, ... ,x_{n+m} are slack variables.

In matrix from these equations can be written as

$$\begin{bmatrix} a_{11} & a_{12} & \dots & a_{1n} & 1 & 0 & \dots & 0 \\ a_{21} & a_{22} & \dots & a_{2n} & 0 & 1 & \dots & 0 \\ \dots & \dots & \dots & \dots & \dots & & \dots & \\ \dots & \dots & \dots & \dots & \dots & & \dots & \\ a_{m1} & a_{m2} & \dots & a_{mn} & 0 & 0 & \dots & 1 \end{bmatrix} \begin{bmatrix} x_1 \\ x_2 \\ \vdots \\ x_n \\ x_{n+1} \\ \vdots \\ x_{n+m} \end{bmatrix}$$

$$= \begin{bmatrix} b_1 \\ b_2 \\ \vdots \\ b_m \end{bmatrix}$$

Taking the initial basis matrix

$B = I_m$, (unit matrix of order m × m).

∴ the initial basic solution is given by

$$x_B = B^{-1} b = I_m\, b = b \geq 0$$

Thus the initial B.F.S. is

$$x_{n+1} = x_{B1} = b_1,\ x_{n+2} = x_{B2} = b_2,\ \ldots\ , x_{n+m} = x_{Bm} = b_m,$$

which can be obtained by writing all the non-basic variables (*i.e.*, given variables) $x_1, x_2, \ldots\ x_n$ equal to zero and solving the equations for the remaining variables (*i.e.*, slack variables) $x_{n+1}, x_{n+2}, \ldots\ , x_{n+m}$.

Case II.

To find initial B.F.S. when all the original constraints have ≥ sign

First we convert all the constraints into equations by inserting surplus variables. The equations obtained are as follows:

$$\begin{aligned} a_{11}\, x_1 + a_{12}\, x_2 + \ldots + a_{1n}\, x_n - x_{n+1} &= b_1 \\ a_{21}\, x_1 + a_{22}\, x_2 + \ldots + a_{2n}\, x_n - x_{n+2} &= b_2 \\ \ldots \quad \ldots \quad \ldots \quad \ldots \quad \ldots \quad \ldots \\ \ldots \quad \ldots \quad \ldots \quad \ldots \quad \ldots \quad \ldots \\ a_{m1}\, x_1 + a_{m2}\, x_2 + \ldots + a_{mn}\, x_n - x_{n+m} &= b_m \end{aligned}$$

Here $x_{n+1}, x_{n+2}, \ldots\ , x_{n+m}$ are surplus variables.

In matrix form, these equations can be writing as

$$\begin{bmatrix} a_{11} & a_{12} & \ldots & a_{1n} & -1 & 0 & \ldots & 0 \\ a_{21} & a_{22} & \ldots & a_{2n} & 0 & -1 & \ldots & 0 \\ \ldots & \ldots & & & \ldots & \ldots & & \ldots \\ \ldots & \ldots & & & \ldots & \ldots & & \ldots \\ a_{m1} & a_{m2} & 0 & 0 & \ldots & & \ldots & -1 \end{bmatrix} \begin{bmatrix} x_1 \\ x_2 \\ \vdots \\ x_n \\ x_{n+1} \\ \vdots \\ x_{n+m} \end{bmatrix} = \begin{bmatrix} b_1 \\ b_2 \\ \vdots \\ b_m \end{bmatrix}$$

If we take the initial basis matrix $B = -I_m$,

we have $x_B = B^{-1} b = = -I_m b = -b \leq 0$, which is not a B.F.S.

To avoid this difficulty we add one more variable to each constraint. These variables are called "**artificial variables**".

Adding surplus and artificial variables, the constraints of the given L.P.P. change to the following equations.

$$a_{11} x_1 + a_{12} x_2 + \ldots + a_{1n} x_n - x_{n+1} + x_{n+m+1} = b_1$$
$$a_{21} x_1 + a_{22} x_2 + \ldots + a_{2n} x_n - x_{n+2} + x_{n+m+2} = b_2$$
$$\ldots \quad \ldots \quad \ldots \quad \ldots \quad \ldots \quad \ldots \quad \ldots$$
$$\ldots \quad \ldots \quad \ldots \quad \ldots \quad \ldots \quad \ldots \quad \ldots$$
$$a_{m1} x_1 + a_{m2} x_2 + \ldots + a_{mn} x_n - x_{n+m} + m_{n+m+m} = b_m$$

Here $x_{n+m+1}, x_{n+m+2}, \ldots, x_{n+m+m}$ are the artificial variables.

In matrix, these equations can be written as

$$\begin{bmatrix} A_{11} & a_{12} & \cdots & a_{1n} & -1 & 0 & \cdots & 0 & 1 & 0 & \cdots & 0 \\ a_{21} & a_{22} & \cdots & a_{2n} & 0 & -1 & \cdots & 0 & 0 & 1 & \cdots & 0 \\ \cdots & \cdots & \cdots & \cdots & \cdots & \cdots & \cdots & \cdots & \cdots & \cdots & \cdots & \cdots \\ \cdots & \cdots & \cdots & \cdots & \cdots & \cdots & \cdots & \cdots & \cdots & \cdots & \cdots & \cdots \\ a_{m1} & a_{m2} & \cdots & a_{mn} & 0 & 0 & \cdots & -1 & 0 & 0 & \cdots & 1 \end{bmatrix}$$

$$\begin{bmatrix} x_1 \\ x_2 \\ \vdots \\ x_n \\ x_{n+1} \\ \vdots \\ x_{n+m} \\ x_{n+m+1} \\ \vdots \\ x_{n+m+m} \end{bmatrix} = \begin{bmatrix} b_1 \\ b_2 \\ \vdots \\ b_m \end{bmatrix}$$

Now taking the basis matrix $B = I_m$.

$\therefore x_B = B^{-1} b = I_m b = b \geq 0$,

which is a basic feasible solution.

$\therefore$ the B.F.S. is $x_{n+m+1} = x_{B1} = b_1$, $x_{n+m+2} = x_{B2} = b_2$, $\ldots$, $x_{n+m+m} = x_{Bm} = b_m$, which can be obtained by writing all the non basic variables (*i.e.*, given variables and surplus variables) $x_1, x_2, \ldots, x_n; x_{n+1}, x_{n+2}, \ldots, x_{n+m}$ equal to zero and solving the equations for the remaining basic variables (*i.e.,* artificial variables) $x_{n+m+1}, x_{n+m+2}, \ldots, x_{n+m+m}$.

Case III.

To find initial B.F.S. when the constraints have '≤', '≥' and '=' signs.

In this case we convert the constraint into equations by inserting slack, surplus and artificial variables. Here the basis matrix $B = I_m$ is obtained by introducing unit column vectors corresponding to the slack and artificial variables.

To obtain the initial B.F.S. we put the non-basic variables equal to zero and solve for remaining basic variables. Here also the initial B.F.S. $x_B = b \geq 0$.

Notes:

1. It is important to note that in all the three cases the initial B.F.S. consists of the constants $b_1, b_2, \ldots, b_m$, (note that $b_1, b_2, \ldots, b_m$ are all positive).
2. Sometimes an identity is present in A without introducing the artificial variables. So there is no need to introduce the artificial variables in such cases.

COMPUTATIONAL PROCEDURE SIMPLEX METHOD

Step 1. If the given problem is of minimization, convert it into the maximization problem.

For this, multiply both sides of the objective function by –1 and put $-Z = Z'$.

If v is the maximum value of Z' then –v will be the minimum value of Z.

Step 2. Make all the bi's non-negative.

The R.H.S. of each of the constraints should be non-negative. If the L.P.P. has a constraint for which a negative bi is given, it should be converted into positive value by multiplying both sides of the constraint by –1.

Step 3. Convert inequalities of constraints into equations.

For this introduce slack or surplus variables. The coefficients of slack or surplus variables in the objective function are zero. Introduce artificial variables, if necessary.

Step 4. Find the initial B.F.S. follow the method discussed in 10 to find the initial B.F.S. If artificial variables are introduced in the L.P.P. then follow the two-method for Big M-method for solving such prob lems.

Step 5. Construct the initial simplex table (starting simplex table).

(See the initial simplex table on page L-101)

It should be remembered that the values of non-basic variables are always zero at each iteration

Step 6. Test the initial B.F.S. for optimality.

Compute the net evaluation Δ_j for each variables x_j by using the formula $\Delta_j = c_j - c_B Y_j$.

Note than in the starting simplex table Δ_j's are same as $-c_j$'s. Also Δ_j's corresponding to basic variables are zero.

Optimality Test

(a) If $\Delta_j \le 0$ for all j, the solution under consideration is **optimal.** Alternative optimal solution will exist if any non-basic Δ_j is also zero.

(b) If $\Delta_j > 0$ for any j *i.e.,* if at least one Δ_j is positive the solution under test is **not optimal**, we must proceed to *improve the solution* (step 7).

(c) If corresponding to maximum positive D_j, all elements of the column Y_j are negative or zero, the solution under test will be **unbounded**.

(d) If optimality condition is satisfied but the value of at least one artificial variables present in the basis is non-zero, the problem will have no **feasible solution**.

Step 7. *Select the entering (incoming) vector and departing (outgoing) vector.*

To improve the initial B.F.S. we select the vector entering the basis matrix and the vector to be removed from the basis matrix.

Incoming Vector. The incoming vector α_k is always selected corresponding to the largest positive value of Δ_j.

If maximum value of Δ_j occurs for more than one α_j then we can select any of these vectors as incoming vector.

Outgoing Vector. The departing vector β_r is selected corresponding to that value of r for which

$$\frac{x_{Br}}{y_{rk}} = \min_i \left\{ \frac{x_{Bi}}{y_{ik}}, y_{ik} > 0 \right\},$$

is α_r is the incoming vector.

Note: If the above mentioned minimum value is not unique then more than one variable will vanish in the next solution. As a result the next solution will be a degenerate B.F.S. for which the outgoing vector is selected in a different way.

Step 8. If α_k is the entering vector and β_r is the outgoing vector then the element yrk which lies at the intersection of minimum ration arrow (←) and incoming vector arrow ↑ is called the pivot element (key element).

We put this element in □.

In order to bring b_r in place of incoming vector Y_k, unity must occupy the place □. In other words key element y_{rk} should be 1. If it is not so then divide all the elements of this row by the key element y_{rk}. Then subtract suitable multiples of the row containing the key element from all other rows and obtain zero at all other positions of this column Y_k. Now bring br in place of α_r and construct the revised simplex table.

Thus we get improve basic feasible solution.

Step 9. Test the improved B.F.S. for optimality.

If it is not optimum, repeat steps 7 and 8 until we obtain an optimums solution.

Example 1:

Maximize $Z = 40x_1 + 35x_2$

subject to $2x_1 + 3x_2 \le 60$

$4x_1 + 2x_2 \le 96,$

$x_1, x_2 \ge 0.$

For the clear understanding of the simplex method the solution to this problem is illustrated below in a step-wise manner.

Solution:

Step 1. The given problem is of maximization and all the b_i's are already positive.

Step 2. Converting the inequalities of constraints into equations by introducing slack variables x_3 and x_4, we get

$$2x_1 + 3x_2 + x_3 = 60 \text{ and } 4x_1 + 3x_2 + x_4 = 96.$$

The coefficients of slack variables are zero in the objective function. Therefore the given problem becomes

$$\text{Maximize } Z = 40x_1 + 35x_2 + 0x_3 + 0x_4$$

subject to $2x_1 + 3x_2 + x_3 + 0x_4 = 60$

$$4x_1 + 3x_2 + 0x_3 + x_4 = 96$$

$$x_1, x_2, x_3, x_4 \geq 0.$$

This is the standardized form of the given problem.

Step 3. The simplex method begins with an initial basic feasible solution in which the values of basic variables are zero.

$\therefore x_1 = 0, x_2 =$ which gives $x_3 = 60, x_4 = 96.$

Step 4. Constructing the initial simplex table.

First Simplex Table

		c_j	**40**	**35**	**0**	**0**	**Min. ratio**
B	c_B	x_B	Y_1	Y_2	Y_3	Y_4	x_B/Y_1
Y_3	0	60	2	3	1	0	60/2 = 30
Y_4	0	96	(4)	3	0	1	96/4 = 24
	←						
$Z=c_B$	$x_B=0$	Δ_j	40 ↑	35	0	0 ↓	

Step 5. Computing Δ_j for all non-basic variables x_j, j = 1, 2, using the formula $\Delta_j = c_j - c_B Y_j$

$$\Delta_1 = c_1 - c_B Y_1 = 40 - (0, 0)(2, 4) = 40$$

$$\Delta_2 = c_2 - c_B Y_2 = 35 - (0, 0)(3, 3) = 35.$$

Since all Δ_j's are not less than or equal to zero therefore the solution is not optimal. So we proceed to next step to improve the solution.

Step 6. Incoming and outgoing vectors.

Since $\Delta_1 = 40$ is the maximum value of Δ_j, j = 1, 2, therefore $\alpha_1 = Y_1$ is the incoming vector.

$$\text{Now } \frac{x_B}{Y_1} = \left(\frac{x_{B1}}{y_{11}}, \frac{x_{B2}}{y_{21}}\right) = \left(\frac{60}{2}, \frac{96}{4}\right) = (30, 24).$$

$$\frac{x_{Br}}{y_{r1}} = \overset{\min}{i}\left[\frac{x_{Bi}}{y_{i1}}, y_{i1} > 0\right] = \overset{\min}{i}\left(\frac{x_{B1}}{y_{11}}, \frac{x_{B2}}{y_{21}}\right) = 24 = \frac{x_{B2}}{y_{21}}$$

$\therefore$ r = 2 which gives $\beta_2 = Y_4$ as the outgoing vector.

$\therefore y_{21} = \alpha_{21}$ is the key element which is equal to 4.

To bring Y_1 in place of Y_4 we proceed in the following manner.

Dividing the second row containing the key element $\alpha_{21} = 4$ by 4 to make if unity.

Now we shall subtract appropriate multiples of this new row form the other remaining row so as to obtain zeros in the remaining positions of Y_1.

Subtitling 2 times of the second row from the first row and subtract 40 times of the second row from the row containing Δ_j's to obtain new values of Δ_j's.

Constructing the second simplex table in which β_2 is replaced by α_1.

Second Simplex Table

B	c_B	c_j / x_B	10 / Y_1	35 / Y_2	0 / Y_3	0 / Y_4	Min ratio x_B/Y_2
Y_3	0	12	0	(3/2)	1	–1/2	8 ←
Y_1	40	24	1	3/4	0	1/4	32
$Z=c_B$	x_B=960	Δ_j	0	5 ↑	0 ↓	–10	

Also computing Δ_j by using the formula $\Delta = c_j - c_B Y_1$ for x_2, x_4, we get

$$\Delta_2 = c_2 - c_B Y_2 = 35 - (0, 40)\left(\frac{3}{2}, \frac{3}{4}\right) = 5$$

$$\Delta_4 = c_4 - c_B Y_4 = 0 - (0, 40)\left(-\frac{1}{2}, \frac{1}{4}\right) = -10.$$

Δ_1 and Δ_3 are zero as they correspond to unit column vectors.

Since all Δ_j's are not less than or equal to zero therefore the solution obtained is not optimal.

$\Delta_2 = 5$ is the maximum value of Δ_j.

$\therefore \alpha_2 = Y_2$ is the incoming vector.

Now $\frac{x_{Br}}{y_{r2}} = \min\limits_{i}\left[\frac{x_{Bi}}{y_{i2}}, y_{i2} > 0\right]$

$= \min\left[\frac{x_{B1}}{y_{12}}, \frac{x_{B2}}{y_{22}}\right] = \min\,[8, 32] = 8$

$= Y_3$

$\alpha_{12} = y_{12} = \frac{3}{2}$ is the key element.

To bring Y_2 in place of Y_3 we proceed in he following manner.

Dividing the first row by 3/2 to make the key element unity.

Subtracting 3/4 times of the first row from the second row and subtracting 5 times of the list row form the two containing Δ_j's to obtain new values of Δ_j's.

Constructing the third simplex table in which Y_3 is replaced by Y_2.

Third Simplex Table

B	c_B	c_j / x_B	40 / Y_1	35 / Y_2	0 / Y_3	0 / Y_4	Min. ratio
Y_2	35	8	0	1	2/3	−1/3	
Y_1	40	18	1	0	−1/2	1/2	
$Z=c_B$	x_B=1000	Δ_j	0	0	−10/3	−25/3	

Since all the Δ_j's are less than or equal to zero therefore the above solution is optimal.

Hence the optimal solution is

$x_1 = 18, x_2 = 8$ and max. $Z = 1000$.

The above solution with its different computational steps can be more conveniently represented by a single table as shown below:

B	c_B	c_j / x_B	40 Y_1	35 Y_2	0 Y_3	0 Y_4	Min. ratio x_B/Y_1
Y_3	0	60	2	3	1	0	60/2 = 30
Y_4	0	96	(40)	3	0	1	96/4 = 24
$Z=c_B$	$x_B=0$	Δ_j	40 ↑	35	0	0 ↓	x_B/y_2
Y_3	0	12	0	(3/2)	1	−1/2	8
Y_1	40	24	1	3/4	0	1/4	32
$Z=c_B$	$x_B=960$	Δ_j	0	5 ↑	0 ↓	−10	
Y_2	35	8	0	1	2/3	−1/3	
Y_1	40	18	1	0	−1/2	1/2	
$Z=c_B$	$x_B=1000$	Δ_j	0	0	−10/3	−25/3	

Example 2:

maximize $Z = 3x_1 + 5x_2 + 4x_3$

subject to $2x_1 + 3x_2 \le 8$

$2x_2 + 5x_3 \le 10$

$3x_1 + 2x_2 + 4x_3 \le 15,\ x_1, x_2, x_3 \ge 0.$

Solution:

The given problem is of maximization and all the b_i's are non-negative.

Converting the inequalities of constraints into equations by introducing slack variables x_4, x_5, x_6. Also the coefficients of slack variables are zero in the objective function. Thus the given problem becomes

$$Z = 3x_1 + 5x_2 + 4x_3 + 0x_4 + 0x_5 + 0x_6$$

subject to $2x_1 + 3x_2 + 0x_3 + x_4 = 8$

$$0x_1 + 2x_2 + 5x_3 + x_5 = 10$$

$$3x_1 + 2x_2 + 4x_3 + x_6 = 15.$$

The starting basic feasible solution is

$x_1 = 0, x_2 = 0, x_3 = 0, x_4 = 8, x_5 = 10, x_6 = 15$. The solution to the problem using simplex method is given in the following table.

B	c_B	c_j / x_B	3 / Y_1	5 / Y_2	4 / Y_3	0 / Y_4	0 / Y_5	0 / Y_6	Min. ratio x_B/Y_2
Y_4	0	8	2	(3)	0	1	0	0	8/3
Y_5	0	10	0	2	5	0	1	0	10/2=5 ←
Y_6	0	15	3	2	4	0	0	1	15/2
$Z=x_B$	$x_B=0$	Δ_j	3	5 ↑	4	0 ↓	0	0	x_B/Y_3
Y_2	5	8/3	2/3	1	0	1/3	0	0	–
Y_5	0	14/3	–4/3	0	(5)	–2/3	1	0	14/15 ←
Y_6	0	29/3	5/3	0	4	–2/3	0	1	29/12
$Z=c_B\ x_B$ =40/3		Δ_j	–1/3	0 ↑	4	–5/3 ↓	0	0	x_B/Y_1
Y_2	5	8/3	2/3	1	0	1/3	0	0	4
Y_3	4	14/15	–4/15	0	1	–2/15	1/5	0	–
Y_6	0	89/15	(41/15)	0	0	–2/15	–4/5	1	89/41 ←
$Z = c_B\ x_B$ = 256/45		Δ_j	11/15 ↑	0	0	–17/15	–4/5	0 ↓	

B	c_B	c_j / x_B	3 / Y_1	5 / Y_2	4 / Y_3	0 / Y_4	0 / Y_5	0 / Y_6	Min. ratio
Y_2	5	50/41	0	1	0	15/41	8/41	–10/41	
Y_3	4	62/41	0	0	1	–6/41	5/41	4/41	
Y_1	3	89/41	1	0	0	–2/41	–12/41	15/41	
$Z = c_B\ x_B$ = 765/41		Δ_j	0	0	0	–45/41	–24/41	–11/41	

In the last table all Δ_j's for non-basic variables are negative therefore it gives optimal solution.

$\therefore$ Optimal solution is

$$x_1 = \frac{89}{41}, x_2 = \frac{50}{41}, x_3 = \frac{62}{41}$$

and $\quad$ Max. $Z = \frac{765}{41}$.

Example 3:

Maximize $Z = 3x_1 + 2x_2 + 5x_3$

subject to $x_1 + 2x_2 + x_3 \leq 430$

$3x_1 + 2x_3 \leq 460$

$x_1 + 4x_2 \leq 420,$

$x_1, x_2, x_3 \geq 0.$

Solution:

The given problem is of maximization and all the b_i's are non-negative.

Convert the inequalities of constraints into equations by introducing slack variables x_4, x_5, x_6.

The given problem becomes

$$Z = 3x_1 + 2x_2 + 5x_3 + 0x_4 + 0x_5 + 0x_6$$

subject to

$$x_1 + 2x_2 + x_3 + x_4 = 430$$

$$3x_1 + 0x_2 + 2x_3 + x_5 = 460$$

$$x_1 + 4x_2 + 0x_3 + x_6 = 420.$$

The initial basic feasible solution is

$$x_1 = 0, x_2 = 0, x_3 = 0, x_4 = 430, x_5 = 460, x_6 = 420.$$

The solution to the problem using simplex algorithm is given in the following table:

Since all Δ_j's are negative or zero therefore the solution is optimal.

Hence the optimal solution is

$$x_1 = 0,\ x_2 = 100,\ x_3 = 230 \text{ and max } Z = 1350.$$

B	c_B	c_j / x_B	3 / Y_1	2 / Y_2	5 / Y_3	0 / Y_4	0 / Y_5	0 / Y_6	Min. ratio / x_B/Y_3
Y_4	0	430	1	2	1	1	0	0	430
Y_5	0	460	3	0	(2)	0	1	0	230 ←
Y_6	0	420	1	4	0	0	0	1	–
$Z=c_B$	$x_B=0$	Δ_j	3	2	5 ↑	0	0 ↓	0	x_B/Y_2
Y_4	0	200	–1/2	(2)	0	1	–1/2	0	100 ←
Y_3	5	230	3/2	0	1	0	1/2	0	–
Y_6	0	420	1	4	0	0	0	1	420/4
Z=1150		Δ_j	–9/2	2 ↑	0	0 ↓	–5/2	0	
Y_2	2	100	–1/4	1	0	1/2	–1/4	0	
Y_3	5	230	3/2	0	1	0	1/2	0	
Y_6	0	20	2	0	0	–2	1	1	
Z=1350	Δ_j	–4	0	0	–1	–2	0		

Example 4:

Minimize $Z = x_2 - 3x_3 + 2x_5$

subject to

$3x_2 - x_3 + 2x_5 \leq 7$

$-2x_2 + 4x_3 \leq 12$

$-4x_2 + 3x_3 + 8x_5 \leq 10,$

$x_2, x_3, x_5 \geq 0.$

Solution:

The given problem is of minimization. Convert it to maximization problem by taking the objective function as Z' = – Z.

The objective function becomes

Max. $Z' = -Z = -x_2 + 3x_3 - 2x_5$.

To convert the inequalities of constraints into equations introducing slack variables x1, x4 and x6, we get

$$x_1 + 3x_2 - x_3 + 0x_4 + 2x_5 + 0x_6 = 7$$
$$0x_1 - 2x_2 + 4x_3 + x_4 + 0x_5 + 0x_6 = 12$$
$$0x_1 - 4x_2 + 3x_3 + 0x_4 + 8x_5 + x_6 = 10$$

and the objective function becomes

$$\text{Max. } Z' = 0x_1 - x_2 + 3x_3 + 0x_4 - 2x_5 + 0x_6.$$

The starting basic feasible solution is

$$x_1 = 7, x_2 = 0, x_3 = 0, x_4 = 12, x_5 = 0, x_6 = 10.$$

The solution to the problem using simplex algorithm is given below:

		c_j	0	–1		0	–20	Min.	ratio
B	c_B	x_B	Y_1	Y_2	Y_3	Y_4	Y_5	Y_6	x_B/Y_3
Y_1	0	7	1	3	–1	0	2	0	–ive
Y_4	0	12	0	–2	(4)	1	0	0	3 ←
Y_6	0	10	0	–4	3	0	8	1	10/3
$Z'=c_B$	$x_B=0$	Δ_j	0	–1	3 ↑	0 ↓	–2	0	x_B/Y_2
Y_1	0	10	1	(5/2)	0	1/4	2	0	4 ←
Y_3	3	3	0	–1/2	1	1/4	0	0	–ive
Y_6	0	1	0	–5/2	0	–3/4	8	1	–ive
Z'=9	Δ_j	0 ↑	1/2 ↓	0	–3/4	–2	0		
Y_2	–1	4	2/5	1	0	1/10	4/5	0	
Y_3	3	5	1/5	0	1	3/10	2/5	0	
Y_6	0	11	1	0	0	–1/2	10	1	
Z' = 11	Δ_j	–1/5	0 ↑	0 ↓	–4/5	–12/5	0		

Since all Δ_j's ≤ 0 therefore the solution given by last table is optimal.

Hence the optimal solution is

$x_2 = 4$, $x_3 = 5$, $x_5 = 0$ and min. $Z = -Z' = -11$.

Example 5:

Maximize $Z = 2x_1 + 5x_2 + 7x_3$

subject to

$3x_1 + 2x_2 + 4x_3 \leq 100$

$x_1 + 4x_2 + 2x_3 \leq 100$

$x_1 + x_2 + 3x_3 \leq 100,$

$x_1, x_2, x_3 \geq 0.$

Solution:

The given problem is of maximization and all b_i's are positive.

Introducing slack variables x_4, x_5, x_6 to convert constraint inequalities into equations, the given problem becomes

$$Z = 2x_1 + 5x_2 + 7x_3 + 0x_4 + 0x_5 + 0x_6$$

subject to $3x_1 + 2x_2 + 4x_3 + x_4 = 100$

$$x_1 + 4x_2 + 2x_3 + x_5 = 100,\ x_1 + x_2 + 3x_3 + x_6 = 100$$

B	c_B	c_j / x_B	2 / Y_1	5 / Y_2	7 / Y_3	0 / Y_4	0 / Y_5	0 / Y_6	Min. ratio x_B/Y_3
Y_4	0	100	3	2	(4)	1	0	0	25 ←
Y_5	0	100	1	4	2	0	1	0	50
Y_6	0	100	1	1	3	0	0	1	100/3
$Z=c_B$	$x_B=0$	Δ_j	2	5	7 ↑	0 ↓	0	0	
Y_3	0	25	3/4	1/2	1	1/4	0	0	50
Y_5	0	50	–1/2	(3)	0	–1/2	1	0	50/3 ←
Y_6	0	25	–5/4	–1/2	0	–3/4	0	1	Neg.
$Z=c_B$	$x_B=175$	Δ_j	–13/4	3/2 ↑	0	–7/4	0 ↓	0	

B	c_B	c_j / x_B	2 / Y_1	5 / Y_2	7 / Y_3	0 / Y_4	0 / Y_5	0 / Y_6	Min. ratio / x_B/Y_3
Y_5	7	50/3	5/6	0	1	1/3	–1/6	0	
Y_2	5	50/3	–1/6	1	0	–1/6	1/3	0	
Y_6	0	100/3	–4/3	0	0	–5/6	1/6	1	
$Z=c_Bx_B$ = 200		Δ_j	–3	0	0	–3/2	–1/2	0	

The starting B.F.S. is

$$x_1 = 0, x_2 = 0, x_3 = 0, x_4 = 100, x_5 = 100, x_6 = 100.$$

The solution to the problem using simplex method is given below:

In the last all Δ_j's are zero or negative therefore the solution is optimal.

Hence the optimal solution is

$$x_1 = 0, x_2 = 50/3, x_3 = 50/3 \text{ and Max. } Z = 200.$$

ARTIFICIAL VARIABLES TECHNIQUE

(a) **Big-M method.** Sometimes in linear programming problems constraints may also have ≥ and = signs after ensuring that all bi ≥ 0. In such problems we first introduce ***surplus variables*** to convert inequalities into equations. In such cases basis matrix is not obtained as an identity matrix in the starting simplex table. To overcome this difficulty we introduce new variables, called, the ***artificial variables*** to each of such constraints. These variables are fictitious and cannot have any physical meaning.

The artificial variables are introduced for the limited purpose of obtaining an ***initial solution.*** It is not relevant whether the objective function is of the maximization or minimization type. Since artificial variables do not represent any quantity relating to the decision problem they must be driven out of the system and must not be present in the final solution. Keeping this in mind a method known as ***big-M method*** or ***Cranes M-method*** was developed. It we suggestedby A. Cranes that a very high penalty can be paid for introducing the artificial variables in the constraints of a given problem by assigning an extremely high negative cost (penalty) to them.

$\therefore$ the objective function can be written as

$$Z = c\,x + 0.x_{slack} + 0.x_{surplus} - M.x_{art}.$$

It is significant to note that the initial solution obtained using the artificial variables is not a feasible solution to the given problem. It only gives the starting point and the artificial variables are driven out in he normal course of applying simplex method to the problem.

Note:

A solution to the problem which does not contain an artificial variables in the basic, represents a feasible solution to the problem.

Example 1:

Maximize $Z = 2x_1 + 4x_2$

subject to $2x_1 + x_2 \leq 18$

$3x_1 + 2x_2 \geq 30$

$x_1 + 2x_2 = 26$

$x_1, x_2 \geq 0.$

Solution:

The problem is of maximization and all b_i's are positive.

Introducing the necessary slack, surplus and artificial variables, the problem reduces to the following form:

$$\text{Max. } Z = 2x_1 + 4x_2 + 0x_3 + 0x_4 - MA_1 - MA_2$$

subject to $2x_1 + x_2 + x_3 = 18$

$3x_1 + 2x_2 + 0x_3 - x_4 + A_1 = 30$

$x_1 + 2x_2 + 0x_3 + 0x_4 + A_2 = 26$

The starting B.F.S. is

$$x_1 = 0,\ x_2 = 0,\ x_3 = 18,\ x_4 = 0,\ A_1 = 30,\ A_2 = 26.$$

The solution to the problem using simplex algorithm is given below:

Computing Δ_j using the formula $\Delta_j = c_j - c_B\, Y_j$, we get

$$\Delta_1 = c_1 - c_B\, Y_1 = 2 - (0, -M, -M)\,(2, 3, 1) = 2 + 4M$$

$$\Delta_2 = c_2 - c_B\, Y_2 = 4 - (0, -M, -M)\,(1, 2, 2) = 4 + 4M$$

$$\Delta_4 = c_4 - c_B\, Y_4 = 0 - (0, -M, -M)\,(0, -1, 0) = -M.$$

First Simplex Table

B	c_B	c_j x_B	2 Y_1	4 Y_2	0 Y_3	0 Y_4	–M A_1	–M A_2	Min. ratio x_B/Y_2
Y_3	0	18	2	1	1	0	0	0	18
A_1	–M	30	3	2	0	–1	1	0	15
A_2	–M	26	1	(2)	0	0	0	1	13 ←
$Z=c_B\,x_B=-56M$	Δ_j		2+4M	4+4M ↑	0	–M	0	0 ↓	

Y_2 is the entering vector. Using min. ratio rule we find that 2 is the pivot element which indicates that A_2 is the outgoing vector.

Second Simplex Table

B	c_B	c_j x_B	2 Y_1	4 Y_2	0 Y_3	0 Y_4	–M A_1	–M A_2	Min. ratio x_B/Y_1
Y_3	0	5	3/2	0	1	0	0	–1/2	10/3
A_1	–M	4	(2)	0	0	–1	1	–1	2 ←
Y_2	4	13	1/2	1	0	0	0	1/2	26
Z = 52 – 4M	Δ_j		2M ↑	0	0	–M	0 ↓	–2 – 2M	

Y_1 is the incoming vector and A_1 is the outgoing vector.

Third Simplex Table

B	c_B	c_j x_B	2 Y_1	4 Y_2	0 Y_3	0 Y_4	–M A_1	–M A_2
Y_3	0	2	0	0	1	3/4	–3/4	1/4
Y_1	2	2	1	0	0	–1/2	1/2	–1/2
Y_2	4	12	0	1	0	1/4	–1/4	3/4
Z=52	Δ_j	0	0	0	0	–M	–M	

Since all Δ_j are negative or zero therefore the solution obtained is optimal.

∴ the optimal solution is

$x_1 = 2, x_2 = 12, x_3 = 2$ and other variables = 0.

The objective function Z = 52.

Example 2:

$$Max.\ Z = x_1 + 2x_2 + 3x_3 - x_4$$

$$subject\ to\ x_1 + 2x_2 + 3x_3 = 15,\ 2x_1 + x_2 + 5x_3 = 20$$

$$x_1 + 2x_2 + x_3 + x_4 = 10,\ x_1, x_2, x_3, x_4 \geq 0.$$

Solution:

The given problem is of maximization and all b_i's are positive. Also the constraints are equations. Examining the constraints we observe that in order to obtain a unit matrix of order 3 we need two more unit vectors as one unit vector is formed by the coefficients of x_4.

Therefore introducing two artificial variables A_1 and A_2 in the first two constraints and assigning large negative penalty –M to the artificial variables, the given problem becomes

$$\text{Max. } Z = x_1 + 2x_2 + 3x_3 - x_4 - MA_1 - MA_2$$

$$\text{subject to } x_1 + 2x_2 + 3x_3 + A_1 = 15$$

$$2x_1 + x_2 + 5x_3 + A_2 = 20,\ x_1 + 2x_2 + x_3 + x_4 = 10.$$

B	c_B	c_j / x_B	1 / Y_1	2 / Y_2	3 / Y_3	–1 / Y_4	–M / A_1	–M / A_2	Min. ratio x_B/Y^3
A^1	–M	15	1	2	3	0	1	0	5
A_2	–M	20	2	1	(5)	0	0	1	4 (Mini) ←
Y^4	–1	10	1	2	1	1	0	0	10
$Z = c_B x_B = -35M - 10$		Δ_j	3M+2	3M+4	8M+4 ↓	0	0	0 ↓	x_B/Y_2
A_1	–M	3	–1/5	(7/5)	0	0	1		15/7 (Mini) ←
Y_3	3	4	2/5	1/5	1	0	0		20
Y_4	–1	6	3/5	9/5	0	1	0		10/3
$Z = c_B x_B = -3M+6$		Δ_j	$\frac{(-M+2)}{5}$	$\frac{7M+16}{5}$ ↑	0	0	0 ↓		x_B/Y_1

Y_2	2	15/7	−1/7	1	0	0	Neg.
Y_3	3	25/7	3/7	0	1	0	25/3
Y_4	−1	15/7	(6/7)	0	0	1	5/2 (Mini) ←
$Z=c_B x_B$ = 90/7		Δ_j	6/7 ↑	0	0	0 ↓	
Y_2	2	5/2	0	1	0	1/6	
Y_3	3	5/2	0	0	1	−1/2	
Y_1	1	5/2	1	0	0	7/6	
$Z=c_B x_B$ = 15		Δ_j	0	0	0	−1	

The starting basic feasible solution is

$$x_1 = 0, x_2 = 0, A_1 = 15, A_2 = 20, x_4 = 10.$$

The solution to the problem using simplex method is given below:

Since all Δ_j's are negative or zero therefore the solution is optimal.

Hence the optimal solution is

$$x_1 = 5/2 = x_2 = x_3, x_4 = 0 \text{ and Max. } Z = 15.$$

Example 3:

Max. $Z = 2x_1 + 3x_2$

subject to $-x_1 + 2x_2 \leq 4$

$x_1 + x_2 \leq 6$

$x_1 + 3x_2 \leq 9$, x_1, x_2 *are unrestricted.*

Solution:

The given problem is of maximization and all b_i's are positive. We know that the simplex method is applicable if the variables are non-negative. In the given problem x_1 and x_2 are unrestricted (may be +, – or zero). *In such cases we replace all these variables by the difference of two non-negative variables and solve the problem in usual manner.*

∴ Making the transformation

$$x_1 = x_1' - x_1'' \text{ and } x_2 = x_2' - x_2'$$

such that $x_1', x_2'', x_2', x_2'' \geq 0$.

The given problem becomes

$$\text{Max. } Z = 2\,(x_1' - x_1'') + 3\,(x_2' - x_2'')$$

$$\text{s.t.} \quad -(x_1' - x_1'') + 2\,(x_2' - x_2'') \le 4$$

$$(x_1' - x_1'') + (x_2' - x_2'') \le 6$$

$$(x_1' - x_1'') + 3\,(x_2' - x_2'') \le 9$$

		c_j	2	−2	3	−3	0	0	0	Min. ratio
B	c_B	x_B	Y_1'	Y_1''	Y_2'	Y_2''	Y_3	Y_4	Y_5	x_B/Y_2'
Y_3	0	4	−1	1	(2)	−2	1	0	0	2 ←
Y^4	0	6	1	−1	1	−1	0	1	0	6
Y^5	0	9	1	−1	3	−3	0	0	1	3
$Z=c^B$ =0	x^B	Δ_j	2	−2	3 ↑	−3	0 ↓	0	0	
Y_2'	3	2	−1/2	1/2	1	−1	1/2	0	0	Neg.
y_4	0	4	3/2	−3/2	0	0	−1/2	1	0	8/3
Y_5	0	3	(5/2)	−5/2	0	0	−3/2	0	1	6/5 ←
$Z=c_B$ =6	x_B	Δ_j	7/2 ↑	−7/2	0	0	−3/2	0	0 ↓	xB/Y3
y_2'	3	13/5	0	0	1	−1	1/5	0	1/5	13
Y_4	0	11/5	0	0	0	0	(2/5)	1	−3/5	11/2 ←
Y_1'	2	6/5	1	−1	0	0	−3/5	0	2/5	Neg.
$Z=c_B$ =51/5	x_B	Δ_j	0	0	0	0	3/5 ↑	0 ↓	−7/5	
Y_2'	3	3/2	0	0	1	−1	0	−1/2	1/2	
Y_3	0	11/2	0	0	0	0	1	5/2	−3/2	
Y_1'	2	9/2	1	−1	0	0	0	3/2	−1/2	
$Z=c_B$ =27/2	x_B	Δ_j	0	0	0	0	−3/2	−1/2		

$$x_1', x_2', x_1'', x_2'' \ge 0.$$

To change constraint inequalities into equations introducing slack variables x_3, x_4 and x_5, the given L.P.P. becomes

$$\text{Max. } Z = 2x_1' - 2x_1'' + 3x_2' - 3x_2'' + 0.x_3 + 0.x_4 + 0.x_5$$

$$-x_1' + x_1'' + 2x_2' - 2x_2'' + x_3 = 4$$

$$x_1' - x_1'' + x_2' - x_2'' + x_4 = 6$$

$$x_1' - x_1'' + 3x_2' - 3x_2'' + x_5 = 9.$$

The starting B.F.S. is $x_3 = 4$, $x_4 = 6$, $x_5 = 9$.

The solution to the problem using simplex algorithm is given below:

Since all Δ_j's are negative or zero therefore the solution is optimal.

$\therefore$ The optimal solution is

$$x_1' = 9/2,\ x_1'' = 0,\ x_2' = 3/2,\ x_2'' = 0,\ \text{Max. } Z = 27/2$$

$$\Rightarrow x_1 = x_1' - x_1'' = 9/2,\ x_2 = x_2' - x_2'' = 3/2,\ \text{Max. } Z = 27/2.$$

Example 4:

Minimize $Z = x_1 + x_2 + x_3$

subject to

$$x_1 - 3x_2 + 4x_3 = 5$$

$$x_1 - 2x_2 \le 3$$

$$2x_2 - x_3 \ge 4,$$

$x_1, x_2 \ge 0$ *and* x_3 *is unrestricted.*

Solution:

First we shall convert the minimization problem into maximization by substituting $Z' = -Z$.

The objective function changes to

$$\text{Max. } Z' = -x_1 - x_1 - x_3.$$

Since x_3 is unrestricted therefore replacing it by

$$x_3' - x_3'' \text{ where } x_3', x_3'' \ge 0.$$

Also introducing slack variables x_4, surplus variables x_5 and artificial variables A_1 and A_2, the given problem becomes

$$\text{Max } Z' = -x_1 - x_2 - x_3' + x_3'' + 0x_4 + 0x_4 + 0x_5 - MA_1 - MA_2$$

s.t. $\quad x_1 - 3x_2 + 4x_3' - 4x_3'' + A_1 = 5$

$x_1 - 2x_2 + x_4 = 3$

$0x_1 + 2x_2 - x_3' + x_3'' - x_5 + A_2 = 4,$

$x_1, x_2, x_3', x_3'', x_4, x_5, A_1, A_2 \geq 0.$

Now solve the problem using simplex algorithm.

(Proceed as in Ex. 3 above).

The optimal solution is

$x_1 = 0, x_2 = 21/5, x_3 = 22/5$ and Min. Z = 43/5.

Example 5:

Apply Big-M method to solve the problem

$Max\ Z = -x_1 - x_2$

subject to

$3x_1 + 2x_2 \geq 30$

$-2x_1 + 3x_2 \leq -30$

$x_1 + x_2 \leq 5,$

$x_1, x_2 \geq 0.$

Solution:

The given problem is of maximization.

Since bi in second constraint is negative therefore multiplying both sides by – 1, we et

$2x_1 - 3x_2 \geq 30.$

To convert inequalities of constraints into equations, introducing slack, surplus and artificial variables wherever necessary the given problem becomes

$Max.\ Z = -x_1 - x_2 + 0x_3 + 0x_4 + 0x_5 \quad MA_1 - MA_2$

subject to

$3x_1 + 2x_2 - x_3 + A_1 = 30$

$2x_1 - 3x_2 - x_4 + A_2 = 30$

$x_1 + x_2 + x_5 = 5$

where $x_1, x_2, x_3, x_4, x_5, A_1, A_2 \geq 0.$

The initial B.F.S. is $x_1 = 0, x_2 = 0, x_3 = 0,$

$x_4 = 0$, $x_5 = 5$, $A_1 = 30$, $A_2 = 30$.

The solution to the problem using simplex method is given below:

B	c_B	c_j / x_B	−1 Y_1	−1 Y_2	0 Y_3	0 Y_4	0 Y_5	−M A_1	−M A_2	Min. ratio x_B/Y_1
A_1	−M	30	3	2	−1	0	0	1	0	10
A_2	−M	30	2	−3	0	−1	0	0	1	15
Y_5	0	5	(1)	1	0	0	1	0	0	5 ←
$Z=c_B\ x_B = -60M$		Δ_j	5M−1 ↑	−M	−M	−M	0 ↓	0	0	
A_1	−M	15	0	−1	−1	0	−3	1	0	
A_2	−M	20	0	−5	0	−1	−2	0	1	
Y_1	−1	5	1	1	0	0	1	0	0	
Z=−35 M−5		Δ_j	0	−6M	−M	−M	−5M	0	0	

Since all $\Delta_j \leq 0$ therefore the solution obtained is optimal. But the artificial variables A_1, A_2 appear in the basis at a positive level which implies that *the given L.P.P. has no feasible solution.*

Note:

Sometimes the constraints may be inconsistent so that there is no feasible solution to the problem. Such a situation is called ***infeasibility***. *In the case of infeasibility one or more artificial variables appear in the final simplex table.*

(b) **Two Phase Method.** As an alternative to the Big-M method, there is another method for dealing with linear programming problems involving artificial variables. This is called the two-dealing with linear and as its name suggests it separates the solution procedure into two phases.

In phase-I, all the he artificial variables are eliminated from the basis.

In phase II, we use the solution from phase I as the initial basic feasible solution and use the simplex method to determine the optimal solution.

COMPUTATIONAL PROCEDURE OF TWO PHASE METHOD

Phase I.

Step 1. To convert each of the constraints into equations, subtract a surplus variable and add an artificial variables.

Step 2. Assign zero coefficients to each of the primary variables and surplus variables and the coefficient (–1) to each of the artificial variables in the objective function. As a result the objective function changes to

$Z = -$ (sum of the artificial variables)

Step 3. Solve the problem formed in step 2 by applying the simplex method. If the original problem has a feasible solution then this problem shall have an optimal solution with optimal value of the objective function equal to zero as each of the artificial variables will be equal to zero.

If max $Z < 0$ and at least one artificial variables appears in the optimum basis at a positive level then the given problem does not possess any feasible solution.

If max $Z = 0$ and at least one artificial variable appears in the optimum basis at zero level, we proceed to phase II.

If max $Z = 0$ and no artificial variable appears in the optimum basis then we proceed to phase II.

Phase II.

In phase II start with the (optimal) solution contained in the final simplex table of the phase I. Assign the actual costs to the variables in the objective function and a zero cost to every surplus variable. Eliminate the artificial which are non-basic at the and of the phase-I. Remove c_j row values of the optimum table and replace them by c_j values of the original problem. Now apply simplex algorithm to the problem contained in the new table to obtain the optimal solution.

Disadvantages of Big-M method over two phase method.

(i) We can always use Big-M method to check the existence of a feasible solution but its computational procedure may be inconvenient because of the manipulation of the constant M. Two phase method eliminates the artificial variables in the beginning.

(ii) When we solve a problem on a digital computer we have to assign some numerical value to M which must be larger than the values $c_1, c_2, \dots$ present in the objective function. But a computer has only a fixed number of digits.

Example 1:

Solve the following problem by using the two-phase method.

$$\text{Min. } Z = 40x_1 + 24x_2$$

subject to $20x_1 + 50x_2 \geq 4800$

$$80x_1 + 50x_2 \geq 7200,$$

$$x_1, x_2 \geq 0.$$

Solution:

The given problem is of minimization. Converting it to maximization by taking the objective function as

$$Z' = -Z.$$

The objective function becomes

$$\text{Max } Z' = -Z = -40x_1 - 24x_2.$$

To change the inequalities of constraint into equations introducing surplus and artificial variables, we get

$$20x_1 + 50x_2 - x_3 + A_1 = 4800$$

$$80x_1 + 50x_1 - x_4 + A_2 = 7200$$

$$x_1, x_2, x_3, x_4, A_1, A_2 \geq 0,$$

x_3, x_4 are surplus variables and A_1, A_2 are artificial variables.

Phase I.

Assigning cost 0 to all other variables and cost –1 to artificial variables, the new objective function of auxiliary problem becomes

$$\text{Max } Z' = 0x_1 + 0x_2 + 0x_3 + 0x_4 - A_1 - A_2$$

subject to the constraints given above.

Now applying the simplex method in the usual manner, we have the following table:

		c_j	0	0	0	0	–1	–1	Min. ratio
B	c_B	x_B	Y_1	Y_2	Y_3	Y_4	A_1	A_2	x_B/Y_1
A_1	–1	4800	20	50	–1	0	1	0	240
A_2	–1	7200	(80)	50	0	–1	0	1	90 ←
$Z=c_B\ x_B$ $= -12000$		Δ_j	100 ↑	100	–1	–1	0	0 ↓	
A_1	–1	3000	0	(75/2)	1	1/4	1	–1/4	80 ←
Y_1	0	90	1	5/8	0	–1/80	0	1/80	144
Z= –2910		Δ_j	0	75/2 ↑	0	1/4	0 ↓	–5/4	
Y_2	0	80	0	1	–2/75	1/50	2/75	–1/50	
Y_1	0	40	1	0	1/60	–1/60	–1/60	–1/60	
Z=0	Δ_j	0	0	0	0	–1	–1		

Since all $\Delta_j \leq 0$ and no artificial variable appears in the basis therefore an optimum solution to the auxiliary problem has been attained.

Phase II.

Now assign the actual costs to the original variables and cost zero to the surplus variables, the objective function becomes

B		c_j	–40	–24	0	0	Min. ratio
B	c_B	x_B	Y_1	Y_2	Y_3	Y_4	x_B/Y_3
x_2	–24	80	0	1	–2/75	1/50	neg.
x_1	–40	40	1	0	(1/60)	–1/60	2400 ←
$Z'=c_B\ x_B= -3520$		Δ_j	0 ↓	0	2/75 ↑	–38/75	
x_2	–24	144	8/5	1	0	–1/50	
x_3	0	2400	60	0	1	–1	
Z'= –3456 Δ_j		–8/5	0	0	–12/25		

Max $Z' = -40x_1 - 24x_2 + 0x_3 + 0x_4$.

Replace the c_j row values in the final simplex table of phase I by the c_j values of the original objective function.

Also delete the artificial variables from the final simplex table of phase I. Now apply the simplex method in the usual manner to this table.

Since all Δ_j's are negative or zero therefore the solution obtained is optimal.

Hence the optimal solution is $x_1 = 0$, $x_2 = 144$ and max $Z = -Z' = 3456$.

Example 2:

Apply two phase simplex method to solve

Min. $Z = x_1 - 2x_2 - 3x_3$

subject to $-2x_1 + x_2 + 3x_3 = 2$

$2x_1 + 3x_2 + 4x_3 = 1,$

$x_1, x_2, x_3 \geq 0.$

Solution:

Converting the objective function to maximization form by substituting $Z = -Z'$, we get

Max. $Z' = -x_1 + 2x_2 + 3x_3$.

Introducing the artificial variables A_1 and A_2, the constraint equations become

$-2x_1 + x_2 + 3x_3 + A_1 = 2$

$2x_1 + 3x_2 + 4x_3 + A_2 = 1$

where $x_1, x_2, x_3, A_1, A_2 \geq 0$.

Phase I.

The new objective function of auxiliary problem is

Max $Z' = 0x_1 + 0x_2 + 0x_3 - A_1 - A_2$,

subject to the constraints mentioned above.

Now apply simplex method in the usual manner to remove artificial variables.

B	c_B	c_j / x_B	0 / Y_1	0 / Y_2	0 / Y_3	–1 / A_1	–1 / A_2	Min. ratio x_B/Y_3
A_1	–1	2	–2	1	3	1	0	2/3
A_2	–1	1	2	3	(4)	0	1	1/4 ←
Z'= –3	Δ_j	0	4	7 ↑	0	0 ↓		
A_1	–1	5/4	–7/2	–5/4	0	1	–3/4	
Y_3	0	1/4	1/2	3/4	1	0	1/4	
Z'= –5/4	Δ_j	–7/4	–5/4	0	0	–3/4		

Since all Δ_j's are negative or zero, an optimum basic feasible solution to the auxiliary problem has been attained. But *max Z' is negative and artificial variable A_1 appears in the basic solution at a positive level therefore the original L.P.P. does not possess any feasible solution.*

L.P.P. HAVING UNBOUNDED SOLUTION

A linear programming problem has unbounded solution if the feasible region is unbounded such that the value of the objective function can be increased indefinitely. It is, however not necessary that an unbounded feasible region should yield an unbounded value for the objective function.

L.P.P. having unbounded feasible region but bounded optimal solution

Example 1:

Max. $Z = 6x_1 - 2x_2$

subject to

$2x_1 - x_2 \leq 2$

$x_1 \leq 4$

$x_1, x_2 \geq 0.$

Solution:

The given problem is of maximization and all b_i's are positive.

To convert inequalities of constraints into equations introducing slack variables x_3 and x_4, given problem becomes

Max. $Z = 6x_1 - 2x_2 + 0x_3 + 0x_4$

subject to

$2x_1 - x_2 + x_3 = 2$

$x_1 + x_4 = 4.$

The starting B.F.S. is

$x_1 = 0,\ x_2 = 0,\ x_3 = 2,\ x_4 = 4.$

The solution to the problem using simplex method is given below:

B	c_B	c_j / x_B	6 / Y_1	–2 / Y_2	0 / Y_3	0 / Y_4	Min. ratio x_B/Y_1
Y_3	0	2	(2)	–1	1	0	1 ←
Y_4	0	4	1	0	0	1	4
$Z=c_B$	$x_B=0$	Δ_j	6 ↑	–2	0 ↓	0	x_B/Y_2
Y_1	6	1	1	–1/2	1/2	0	Neg.
Y_4	0	3	0	(1/2)	–1/2	1	6 ←
Z=6	Δ_j	0	1 ↑	–3	0 ↓		
Y_1	6	4	1	0	0	1	
Y_2	–2	6	0	1	–1	2	
Z=12	Δ_j	0	0	–2	–2		

Since all Δ_j's are negative or zero therefore the solution obtained is optimal.

The optimal solution is

$x_1 = 4,\ x_2 = 6$ and

max $Z = 12.$

From the first simplex table we observe that the elements of column of Y_2 are either negative or zero which indicates that the feasible region is not bounded.

Hence a linear programming problem having unbounded feasible solution may have a bounded optimal solution.

Example 2:

Max. $Z = 10x_1 + 20x_2$

subject to $2x_1 + 4x_2$

$x_1 + 5x_2 \geq 15,$

$x_1, x_2 \geq 0.$

Solution:

The given problem is of maximization and all b_i's are positive.

Introducing surplus variables x_3, x_4 and artificial variables A_1, A_2, the given problem becomes

Max. $Z = 10x_1 + 20x_2 + 0x_3 + 0x_4 - MA_1 - MA_2$

subject to $2x_1 + 4x_2 - x_3 + A_1 = 16$

$x_1 + 5x_2 - x_4 + A_2 = 15,$

$x_1, x_2, x_3, x_4, A_1, A_2, \geq 0.$

The starting B.F.S. is

$x_1 = 0, x_2 = 0, x_3 = 0, x_4 = 0, A_1 = 16, A_2 = 15.$

The final simplex table in respect of this problem is

		c_j	10	20	0	0	–M	–M	Min. ratio
B	c_B	x_B	Y_1	Y_2	Y_3	Y_4	A_1	A_2	
Y_1	10	5	1	5	0	–1	0	1	
Y_3	0	14	0	6	1	–2	–1	2	
Z=150	Δ_j	0	–30	0	10 ↑	–M	–M – 10		

From the table we observe that Y_4 is the incoming vector but we can not find outgoing vector because all min. ratios are negative. Therefore the solution is unbounded.

Note: If in a situation there is at least one Δ_j greater than zero but there is no non-negative ratios or min. ratios $\rightarrow \infty$ then we have unbounded solution.

L.P.P. having more than one optimum solution.

Example 3:

Max. $Z = 6x_1 + 4x_2$

subject to $2x_1 + 3x_2 \leq 30$

$3x_1 + 2x_2 \leq 30$

$x_1 + x_2 \geq 3,\ x_1,\ x_2 \geq 0.$

Solution:

To convert inequalities into equations, introducing slack, surplus and artificial variables, the problem becomes

Max. $Z = 6x_1 + 4x_2 + 0x_3 + 0x_4 + 0x_5 - MA_1$

		c_j	6	4	0	0	0	–M	Min. ratio
B	c_B	x_B	Y_1	Y_2	Y_3	Y_4	Y_5	A_1	x_B/Y_1
Y_3	0	30	2	3	1	0	0	0	15
Y_4	0	24	3	2	0	1	0	0	8
A_1	–M	3	(1)	1	0	0	–1	1	3 ←
$Z=c_B\ x_B = -3M$		Δ_j ☐	6+M	4+M ↑	0	0	–M	0 ↓	x_B/Y_5
Y_3	0	24	0	1	1	0	2		12
Y_4	0	15	0	–1	0	1	(3)		5 ←
Y_1	6	3	1	1	0	0	–1		–
Z=18		Δ_j	0	–2	0 ☐	0	6 ↓	↑	
Y_3	0	14	0	(5/3)	1	–2/3	0		
Y_5	0	5	0	–1/3	0	1/3	1		
Y_1	6	8	1	2/3	0	1/3	0		
Z=48		Δ_j	0	0	0	–2	0		

subject to

$$2x_1 + 3x_2 + x_3 = 30$$
$$3x_1 + 2x_2 + x_4 = 24$$
$$x_1 + x_2 - x_5 + A_1 = 3,$$
$$x_1, x_2, x_3, x_4, x_5, A_1 \geq 0.$$

The starting B.F.S. is

$$x_1 = 0, x_2 = 0, x_3 = 30, x_4 = 24, x_5 = 0, A_1 = 3.$$

The solution to the problem using simplex method is given below:

Since all $\Delta_j \leq 0$ therefore the solution is optimal. The optimal solution is $x_1 = 8$, $x_2 = 0$, max $Z = 48$.

Again $\Delta_2 = 0$ and Y_2 is not in the basis B therefore an alternative solution also exists. Thus the problem does not have unique solution.

Taking Y_2 as incoming and Y_3 as outgoing vector we obtain the following simplex table:

B	c_B	c_j x_B	6 Y_1	4 Y_2	0 Y_3	0 Y_4	0 Y_5	Min. ratio
Y_2	4	42/5	0	1	3/5	–2/5	0	
Y_5	0	39/5	0	0	1/5	1/5	1	
Y_1	6	12/5	1	0	–2/5	3/5	0	
$Z = c_B x_B \Delta_j$ =48		0	0	0	–2	0		

All $\Delta_j \leq 0$ therefore this solution is also optimal having the same maximum value of Z.

Second optimal solution is

$$x_1 = 12/5, \; x_2 = 42/5 \text{ and}$$

max $Z = 48$.

We know that the convex combination of B.F.S. is also an optimal solution.

Thus if we obtain two alternative optimum solution then we can obtain any number of optimum solutions.

For any arbitrary value of λ the following table gives different optimum solution:

Variables	I Sol.	II Sol.	Gen. Sol.
x_1	8	12/5	$8\lambda + (12/5)(1 - \lambda)$
x_2	0	42/5	$0\lambda + (42/5)(1 - \lambda)$
x_3	14	0	$14\lambda + 0(1 - \lambda)$
x_4	0	0	$0\lambda + (1 - \lambda)$
x_5	5	39/5	$5\lambda + (39/5)(1 - \lambda)$
A_1	0	0	$0\lambda + 0(1 - \lambda)$

Taking a particular value of λ, $\lambda = \frac{1}{2}$ (say),

the third optimal solution is

$$x_1 = 26/5,\ x_2 = 21/5 \text{ and Max. } Z = 48.$$

SOLUTION OF SIMULTANEOUS LINEAR EQUATIONS BY SIMPLEX METHOD

To solve a system of n simultaneous linear equations by simplex method, a dummy objective function is introduced in which every given variable is assigned 0 (zero) cost and every artificial variable is assigned cost (–1).

The reformulated linear programming problem is then solved by using simplex method.

Example 1:

Solve the following system of linear equations by using simplex method

$$x_1 - x_3 + 4x_4 = 3$$

$$2x_1 - x_2 = 3$$

$$3x_1 - 2x_2 - x_4 = 1$$

$$x_1, x_2, x_3, x_4 \geq 0.$$

Solution:

Introducing the dummy objective function Z with costs zero to each given variable and cost (–1) to each artificial variable the given equations can be written as a linear programming problem in the following form

Max. $Z = 0x_1 + 0x_2 + 0x_3 + 0x_4 - A_1 - A_2 - A_3$

subject to

$x_1 - x_3 + 4x_4 + A_1 = 3$

$2x_1 - x_2 + A_2 = 3$

$3x_1 - 2x_2 - x_4 + A_3 = 1,$

$x_1, x_2, x_3, x_4, A_1, A_2, A_3 \geq 0.$

The starting B.F.S. is

$x_1 = 0, x_2 = 0, x_3 = 0,$

$x_4 = 0, A_1 = 3, A_2 = 3, A_3 = 1.$

The solution to the problem using simplex algorithm is

		c_j	0	0	0	0	–1	–1	–1	**Min. ratio**
B	c_B	x_B	Y_1	Y_2	Y_3	Y_4	A_1	A_2	A_3	x_B/Y_1
A_1	–1	3	1	0	–1	4	1	0	0	3
A_2	–1	3		–1	0	0	0	1	0	3/2
A_3	–1	1	(3)	–2	0	–1	0	0	1	1/3
										←
$Z = c_B x_B$		Δ_j	6	–3	–1	3	0	0	0	X_B/Y_4
= –7			↑						↓	
A_1	–1	8/3	0	2/3	–1	(13/3)	1	0		8/13
A_2	–1	7/3	0	1/3	0	2/3	0	1		7/2
Y_1	0	1/3	1	–2/3	0	–1/3	0	0		Neg.
$Z = c_B x_B$		Δ_j	0	1	–1	5	0	0		
= – 5						↑	↓			
Y_4	0	8/13	0	(2/13)	–3/13	1		0		4
										←
A_2	–1	25/13	0	3/13	2/13	0		1		25/3
Y_1	0	7/13	1	–8/13	–1/13	0		0		Neg.
–25/13		Δ_j	1	3/13	2/13	0		0		x_B/Y_3
			↑		↓					

Y_2	0	4	0	1	–3/2	13/2	0	Net.
A_2	–1	1	0	0	(1/2)	–3/2	1	2 ←
Y_1	0	3	1	0	–1	4	0	Neg.
Z= –1		Δ_j	0	0	1/2 ↑	–3/2	0 ↓	
Y_2	0	7	0	1	0	2		
Y_3	0	2	0	0	1	–3		
Y_1	0	5	1	0	0	1		
Z=0	Δ_j	0	0	0	0			

Since all $\Delta_j \leq 0$ therefore the problem has optimal solution given by

$x_1 = 5, x_2 = 7,$

$x_3 = 2, x_4 = 0.$

INVERSE OF A MATRIX BY SIMPLEX METHOD

Consider any n × n non-singular real matrix A.

Also consider a system of equation $A x = b, x \geq 0$,

where b is a dumbly real matrix of order n × 1.

Introduce the artificial variables $x_a \geq 0$ and a dummy objective function Z with cost zero to variables in x and cost (–1) to each artificial variable. Now find the solution of the following linear programming problem using simplex method

Max. $Z = 0.\, x - 1.x_{art}$

subject to $A x = b,\ x, x_a \geq 0$.

In the last iteration when the column of A become the columns of unit matrix I, the inverse of matrix A is given by those column vectors of the table which were the column of the initial basis.

Example:

Apply simplex method to find the inverse of the matrix $\begin{bmatrix} 4 & 3 \\ 3 & 2 \end{bmatrix}$.

Solution:

Let $A = \begin{bmatrix} 4 & 3 \\ 3 & 2 \end{bmatrix}$ and

$b = \begin{bmatrix} 4 \\ 6 \end{bmatrix}$ be a dummy real vector.

Consider the system of equations

$$A\,x = b \text{ given by } \begin{bmatrix} 4 & 3 \\ 3 & 2 \end{bmatrix}\begin{bmatrix} x_1 \\ x_2 \end{bmatrix}$$

$$= \begin{bmatrix} 4 \\ 6 \end{bmatrix},\ x_1, x_2 \geq 0.$$

Introducing the artificial variables x_3, x_4 and the dummy objective function Z, the linear programming problem becomes

$$\text{Max. } Z = 0x_1 + 0x_2 - x_3 - x_4$$

$$\text{s.t.} \quad 4x_1 + 3x_2 + x_3 = 4$$

$$3x_1 + 2x_2 + x_4 = 6,$$

$$x_1, x_2, x_3, x_4 \geq 0.$$

The starting B.F.S. is $x_3 = 4$, $x_4 = 6$.

B	c_B	c_j / x_B	**0** / Y_1	**0** / Y_2	**–1** / A_1	**–1** / A_2	**Min. ratio** x_B/Y_1
A_1	–1	4	(4)	3	1	0	4/4 = 1 ←
A_2	–1	6	3	2	0	1	6/3 = 2
$Z = c_B\, x_B = -10$		Δ_j	7 ↑	5	0 ↓	0	
Y_1	0	1	1	3/4	1/4	0	
A_2	–1	3	0	(1/4)	–3/4	1	
$Z = c_B\, x_B = -3$		Δ_j	0	–1/4 ↑	–7/4	0	

Since all Δ_j's are negative or zero therefore the solution is optimal. But the matrix A is still not converted into the unit matrix. For this introduce Y_2 into the basis and drop A_2. The new simplex table becomes

B	c_B	c_j / x_B	0 / Y_1	0 / Y_2	–1 / A_1	–1 / A_2
Y_1	0	10	1	0	–2	3
Y_2	0	–12	0	1	3	–4
$Z=c_B$	$x_B=0$	Δ_j	0	0	–1	–1

Since the columns of the given matrix A have become the columns of the unit matrix I therefore

$$A^{-1} = \begin{bmatrix} -2 & 3 \\ 3 & -4 \end{bmatrix}.$$

TO COMPUTE THE INVERSE OF A MATRIX WHOSE ONE COLUMN IS DIFFERENT FROM THAT OF A MATRIX WHOSE INVERSE IS KNOWN

$B = (\beta_1, \beta_2, \ldots, \beta_n)$ is the given non-singular matrix whose inverse is known.

Consider a matrix

$$B_\alpha = (\beta_1, \beta_2, \ldots, \beta_{r-1}, \alpha, \beta_{r+1}, \ldots, \beta_n).$$

We have obtained the matrix Ba by replacing the rth column of matrix B by the column vector a.

If we have $\alpha = \sum_{i=1}^{n} y_i \beta_i$,

Ba will be non-singular iff $y_r \neq 0$.

Using components of the vector $Y = B^{-1}\alpha$, we introduce a new matrix E where E is a square matrix of order n which differs from identity matrix In the rth column only. To obtain the rth column of matrix E we proceed in the following manner:

We have $\alpha = y_1 \beta_1 + y_2 \beta_2 + \ldots + y_r \beta_r + \ldots + y_m \beta_m$.

Transposing yr br to the left and a to the right, we get

$$-y_r \beta_r = y_1 \beta_1 + y_2 \beta_2 + \ldots + (-\alpha) + \ldots + y_n \beta_n$$

$$\text{or } \beta_r = -\frac{y_1}{y_r}\beta_1 - \frac{y_2}{y_r}\beta_2 - \ldots + \frac{1}{y_r}\alpha - \ldots - \frac{y_n}{y_r}\beta_n$$

$$= \frac{y_1}{y_r}\beta_1 - \frac{y_2}{y_r}\beta_2 - \ldots + \frac{1}{y_r}\alpha - \ldots - \frac{y_n}{y_r}\beta_n \; (\beta_1, \beta_2, \ldots, \alpha, \ldots \beta_n)$$

$= A_r B_\alpha,$

where $A_r = \left(-\frac{y_1}{y_r}, -\frac{y_2}{y_r}, \ldots, \frac{1}{y_r}, -\frac{y_n}{y_r}\right)$ gives the rth column of E.

Now $B_\alpha E = B_\alpha (e_1, e_2, \ldots, e_{r-1}, A_r, e_{r+1}, \ldots, e_n)$

$= (B_\alpha e_1, B_\alpha e_2, \ldots, B_\alpha e_{r-1}, B_\alpha A_r, \ldots, B_\alpha e_n)$

$= (\beta_1, \beta_2, \ldots, \beta_{r-1}, \beta_r, \beta_{r+1}, \ldots, \beta_n),$

since $B_\alpha e_1 = (\beta_1, \beta_2, \ldots, \alpha, \ldots, \beta_n)(1, 0, 0, \ldots, 0)$

$= \beta_1 + 0 + 0 + \ldots + 0 = \beta_1$ etc.

$= B.$

Since $B = B\alpha E$

$\therefore BB^{-1} = B_\alpha EB^{-1}$

$\Rightarrow I = B_\alpha EB^{-1}$

$\Rightarrow B_\alpha^{-1} I = (B_\alpha^{-1} B_\alpha) EB^{-1}$

$\Rightarrow B_\alpha^{-1} = IEB^{-1}$

$= EB^{-1}.$

Example:

Find the inverse of the matrix

$B_\alpha = (\beta_1, \beta_2, \alpha)$

where $B = (\beta_1, \beta_2, \beta_3)$,

$$\alpha = \begin{bmatrix} 2 \\ 1 \\ 3 \end{bmatrix} \textit{ and } B^{-1} = \begin{bmatrix} 7 & -3 & -3 \\ -1 & 1 & 0 \\ -1 & 0 & 1 \end{bmatrix}.$$

Solution:

We know that $Y = (y_1, y_2, y_3) = B^{-1}\alpha$

$$\therefore Y = \begin{bmatrix} 7 & -3 & -3 \\ -1 & 1 & 0 \\ -1 & 0 & 1 \end{bmatrix} \begin{bmatrix} 2 \\ 1 \\ 3 \end{bmatrix} = \begin{bmatrix} 2 \\ -1 \\ 1 \end{bmatrix}$$

or $\quad y_1 = 2, y_2 = -1, y_3 = 1.$

$$\text{Now } A_3 = \left(-\frac{y_1}{y_3}, -\frac{y_2}{y_3}, \frac{1}{y_3}\right) = (-2, 1, 1)$$

$$\therefore E = (e_1, e_2, A_3) = \begin{bmatrix} 1 & 0 & -2 \\ 0 & 1 & 1 \\ 0 & 0 & 1 \end{bmatrix}$$

$$\text{Now} \quad B_\alpha^{-1} = EB^{-1} = \begin{bmatrix} 1 & 0 & -2 \\ 0 & 1 & 1 \\ 0 & 0 & 1 \end{bmatrix} \begin{bmatrix} 7 & -3 & -3 \\ -1 & 1 & 0 \\ -1 & 0 & 1 \end{bmatrix}$$

$$= \begin{bmatrix} 9 & -3 & -5 \\ -2 & 1 & 1 \\ -1 & 0 & 1 \end{bmatrix}.$$

SOME DEFINITIONS AND NOTATIONS

Here we shall introduce some notations and definitions which are extremely useful in the discussion of simplex algorithm.

Consider a linear programming problem which after introducing slack and surplus variables is as follows:

$$\text{Max. } Z = c_1 x_1 + c_2 x_2 + \ldots + c_n x_n + 0x_{n+1} + 0x_{n+2} + \ldots + 0x_{n+m}$$

subject to $a_{11} x_1 + a_{12} x_2 + \ldots + a_{1j} x_j + \ldots + a_{1n} x_n + x_{n+1} = b_1$

$a_{21} x_1 + a_{22} x_2 + \ldots + a_{2j} x_j + \ldots + a_{2n} x_n + x_{n+2} = b_2$

... ...

... ...

$a_{m1} x_1 + a_{m2} x_2 + \ldots + a_{mj} x_j + \ldots + a_{mn} x_n + x_{n+m} = b_m,$

$x_i \geq 0$ for all $i = 1, 2, \ldots, N$, where $N = n + m$

$b_1, b_2, \ldots, bm$ are all positive.

The above linear programming problem can be easily converted to matrix form

Max. $Z = c\,x$

subject to $A\,x = b,\ x \geq 0$

where $A = [a_{ij}]_{m \times N}$ is the coefficient matrix of order $m \times N$,

$x = [x_1, x_2, x_3, \ldots, x_n, \ldots, x_N],$

$c = [c_1, c_2, \dots, c_n, 0, 0, \dots, 0]_{1\times N}$

and $\quad b = [b_1, b_2, \dots, b_m]$,

where x and b are column vectors of order $N \times 1$ and $m \times 1$ respectively.

For convenience, we shall represent column vectors by row vectors without using transpose symbol. There should be no confusion in understanding scalar multiplication of two vectors c and x.

If we denote the jth column of the matrix A by a_j, $j = 1, 2, \dots, N$, we can write

$$A = [\alpha_1, \alpha_2, \dots, \alpha_N].$$

Form a non-singular sub-matrix B of order $m \times m$ whose column vectors are m number of linearly independent columns selected from A. If we denote these columns by $\beta_1, \beta_2, \dots, \beta_m$ then

$$B = [\beta_1, \beta_2, \dots, \beta_m].$$

B is called the ***basis matrix***.

Since the vectors $\beta_1, \beta_2, \dots \beta_m$ are linearly independent they form a basis for E^m.

Therefore each $\alpha_j \in A \subset E^m$ can be expressed as a linear combination of vectors of B. Thus we can write

$$\alpha_j = \beta_1 y_{1j} + \beta_2 y_{2j} + \dots + \beta_m y_{mj}$$

$$= (\beta_1, \beta_2, \dots, \beta_m) \begin{bmatrix} y_{1j} \\ y_{2j} \\ \vdots \\ y_{mj} \end{bmatrix}$$

or $\quad \alpha_j = BY_j$, where $Y_j = \begin{bmatrix} y_{1j} \\ y_{2j} \\ \vdots \\ y_{mj} \end{bmatrix}_{m\times 1}$,

$y_{1j}, y_{2j}, \dots, y_{mj}$ are the scalars required to express α_j in such a form.

Also $\alpha_j = BY \Rightarrow Y_j = B^{-1} a_j$.

The vector Yj will change if the columns of A forming B change.

Any basis matrix B will provide a *basic solution* to $A x = b$.

The variables corresponding to $\beta_1, \beta_2, \ldots, \beta_m$ are denoted by $x_{B1}, x_{B2}, \ldots, x_{Bm}$ respectively and are called ***basis variables.***

We denote the column vector of these m basic variables by x_B

$\therefore x_B = [x_{B1}, x_{B2}, \ldots, x_{Bm}]$

and $\quad x_B = B^{-1} b.$

This is called the basic feasible solution (B.F.S.) of the L.P.P.

Since $x_{B1}, x_{B2}, \ldots, x_{Bm}$ are the basic variables, the remaining n variables belonging to x are called non-basic variables.

We shall denote the coefficients of basic variables $x_{B1}, x_{B2}, \ldots, x_{Bm}$ in the objective function Z by $c_{B1}, c_{B2}, \ldots, c_{Bm}$.

Corresponding to any x_B, c_B will represent the row vector contain the constants $c_{B1}, c_{B2}, \ldots, c_{Bm}$.

$\therefore c_B = (c_{B1}, c_{B2}, \ldots, c_{Bm}).$

Since for any basic feasible solution all non-basic variables are zero therefore the objective function Z becomes

$$Z = c_{B1} x_{B1} + c_{B2} x_{B2} + \ldots + c_{Bm} x_{Bm} + 0$$

$$= (c_{B1}, c_{B2}, \ldots, c_{Bm}) (x_{B1}, x_{B2}, \ldots, x_{Bm})$$

$$= c^B x_B.$$

Finally, we introduce a new variables Z_j, given by

$$Z_j = y_j c_{B1} + y_{2j} c_{B2} + \ldots + y_{mj} c_{Bm}$$

$$= (c_{B1}, c_{B2}, \ldots c_{Bm}) (y_{1j}, y_{2j}, \ldots, y_{mj})$$

$$= c_B y_j.$$

Therefore exists Z_j for each α_j and it changes as the columns of A forming B change.

FUNDAMENTAL THEOREM OF LINEAR PROGRAMMING

Theorem:

If a linear programming problem

max $Z = c x$ *subject to* $A x = b, x \geq 0,$

where A is an $m \times N$, $(N = m + n)$ *matrix of coefficients given by* $A = (\alpha_1, \alpha_2, \ldots, \alpha_N)$ *has an optimal feasible solution then at least one basic feasible solution must be optimal.*

Proof:

Let $x^* = [x_1, x_2, \ldots, x_N]$

be an optimal feasible solution of the given linear programming problem and

$$Z^* = \sum_{i=1}^{N} c_i x_i$$

be the corresponding optimum value of the objective function.

Also suppose that k ($k \leq N$) variables in x^* are non-zero and the remaining $N - k$ variables are zero. For the sake of convenience we can assume that the first k variables of x^* are non-zero.

Thus $x^* = (x_1, x_2, \ldots, x_k, 0, 0, \ldots, 0)$.

$$\therefore \quad \sum_{i=1}^{k} x_i \alpha_i = b \qquad \ldots(1)$$

and

$$Z^* = \sum_{i=1}^{k} c_i x_i . \qquad \ldots(2)$$

Now two possibilities may arise:

The vectors $\alpha_1, \alpha_2, \ldots, \alpha_k$ may be either linearly independent or dependent.

Case 1.

If $\alpha_1, \alpha_2, \ldots, \alpha_k$ are linearly independent.

Any feasible solution for which the vectors a_i, $i = 1, \ldots, k$ associated with non-zero variables x_i, $i = 1, \ldots, k$ are L.I. is called a basic feasible solution. Thus x^* is a B.F.S. which is also optimal.

Hence the result of the theorem is true.

The solution x^* is degenerate if $k < m$ and non-degenerate if $k = m$.

Case 2.

If $\alpha_1, \alpha_2, \ldots, \alpha_k$ are linearly dependent and $k > m$.

In this case we shall reduce the number of non-zero variables step by step by step until we obtain a B.F.S. from feasible solution.

Since $\alpha_1, \alpha_2, \ldots, \alpha_k$ are linearly dependent therefore there exist scalars $\lambda_1, \alpha_2, \ldots, \lambda_k$ such that

$$\lambda_1 \alpha_1 + \lambda_2 \alpha_2 + \ldots + \lambda_k \alpha_k = 0$$

or $$\sum_{i=1}^{k} \lambda_i \alpha_i = 0 \qquad \ldots(3)$$

and at least one $\lambda_i \neq 0$.

We can assume that at least one li is positive because if none is positive then we can multiply the equation (3) by –1 and get positive li.

Now, suppose

$$v = \max_{1 \le i \le k} \left(\frac{\lambda_i}{x_i} \right). \qquad \ldots(4)$$

Obviously v is positive since $x_i \geq 0$, i = 1, 2, ..., k and at least one λ_i is positive.

Multiplying both sides of (3) by $\frac{1}{v}$ and subtracting from (1), we get

$$\sum_{i=1}^{k} \left(x_i - \frac{\lambda_i}{v} \right) \alpha_i = b. \qquad \ldots(5)$$

(5) gives

$$x' = \left[x_1 - \frac{\lambda_1}{v}, x_2 - \frac{\lambda_2}{v}, \ldots, x_k - \frac{\lambda_k}{v}, 0, 0, \ldots, 0 \right]$$

a new solution of the matrix equation A x = b.

Also from (4) for i = 1, 2, ... ,k, we have

$$v \geq \frac{\lambda_i}{x_i} \text{ or } x_i \geq \frac{\lambda_i}{v}$$

or $$x_i - \frac{\lambda_i}{v} \geq 0.$$

Since all the components of x' are non-negative therefore x' is a feasible solution to the given L.P.P.

Also $v = \frac{\lambda_i}{x_i}$, for at least one i

i.e., $v - \frac{\lambda_i}{x_i} = 0$ therefore the new feasible solution x' cannot have more than k – 1 non-zero variables.

In this way we have derived a new feasible solution from the given optimal feasible solution which contains less number of non-zero variables. This solution is B.F.S. if the column vectors associated to non-zero variables in this new solution are L.I. If these associated vectors are not L.I., we shall repeat the whole reduction procedure as explained above. Continuing in this way for finite number of times we will derive a solution in which columns corresponding to positive variables are L.I. *i.e.,* we will obtain a B.F.S. of the system.

Now it remains to prove that x' is also an optimal solution.

Let Z' be the new value of the objective function corresponding to the new solution x'.

$$\text{Then } Z' = \sum_{i=1}^{k} c_i \left(x_i - \frac{\lambda_i}{v} \right) = \sum_{i=1}^{k} c_i x_i - \frac{1}{v} \sum_{i=1}^{1} c_i \lambda_i$$

or $$Z' = Z^* - \frac{1}{v} \sum_{i=1}^{k} c_i \lambda_i, \text{ from (2)} \qquad ...(6)$$

Now for optimality, we must have

$$Z' = Z^*$$

i.e., x' will be optimal solution only if

$$\sum_{i=1}^{k} c_i \lambda_i = 0. \qquad ...(7)$$

We shall prove this result by contradiction.

If possible let us assume that

$$\sum_{i=1}^{k} c_i \lambda_i \neq 0.$$

$$\text{Then either } \sum_{i=1}^{k} c_i \lambda_i < 0 \qquad ...(A)$$

$$\Rightarrow \sum_{i=1}^{k} c_i \lambda_i > 0. \qquad ...(B)$$

In either of these two cases, we can find a real number r such that

$r \sum_{i=1}^{k} c_i \lambda_i > 0$ [in case (A) r will be negative and in case (B) r will be positive]

$$\Rightarrow \sum_{i=1}^{k} c_i (r\lambda_i) > 0$$

$$\Rightarrow \sum_{i=1}^{k} c_i (r\lambda_i) + \sum_{i=1}^{k} c_i x_i > \sum_{i=1}^{k} c_i x_i$$

$$\Rightarrow \sum_{i=1}^{k} c_i (x_i + r\lambda_i) > Z^*, \text{ from (2).} \quad \text{...(8)}$$

Multiplying equation (3) by r and adding to (1), we get

$$\sum_{i=1}^{k} x_i \alpha_i + \sum_{i=1}^{k} r\lambda_i \alpha_i = b$$

$$\Rightarrow \sum_{i=1}^{k} (x_i + r\lambda_i)\alpha_i = b.$$

Obviously $[x_1 + r\lambda_1, x_2 + r\lambda_2, \ldots, x_k + r\lambda_k, 0, 0, \ldots, 0]$ is also a solution of the matrix equation A x = b for all values of r.

Furthermore, we can choose r in infinitely many ways for which the above solution also satisfies the non-negative restrictions.

Let us choose r such that

$$x_i + r\lambda_i \geq 0, \ i = 1, \ldots, k$$

or $\quad r\lambda_i \geq -x_i$

$$\therefore r \geq -\frac{x_i}{\lambda_i} \text{ if } \lambda_i > 0,$$

$$r \leq -\frac{x_i}{\lambda_i} \text{ if } \lambda_i < 0$$

and r is unrestricted if $\lambda_i = 0$.

Thus $x_i + r\lambda_i \geq 0$ for i = 1, 2, ... ,k if we select r such that

$$\max_{\substack{i \\ (\lambda_i > 0)}} \left(-\frac{x_i}{\lambda_i}\right) \leq r \leq \min_{\substack{i \\ (\lambda_i < 0)}} \left(-\frac{x_i}{\lambda_i}\right) \quad \text{...(9)}$$

which means that interval given by (9) is non-empty.

Thus r lies in the non-empty interval given in (9).

Consequently, an infinite number of solution given by

$$[x_1 + r\lambda_1, x_2 + r\lambda_2, ..., x_k + x_k + r\lambda_k, 0, 0, ..., 0]$$

satisfy the non-negative restrictions as well.

Now, returning to result (8) we king that $\sum_{i=1}^{k} c_i (c_i + r\lambda_i)$ yields the value of the objective function which is strictly greater than the greatest value (or optimal value) Z* of objective function, which is impossible.

$\therefore$ we must have $\sum_{i=1}^{k} c_i \lambda_i = 0$

or $Z' = Z^*$.

Hence $x' = \left[x_i - \frac{\lambda_1}{v}, x_2 - \frac{\lambda_2}{v}, \ldots, x_k - \frac{\lambda_k}{v}, 0, 0, \ldots, 0\right]$

is also an optimal solution.

This proves the theorem.

REDUCTION OF FEASIBLE SOLUTION TO BASIC FEASIBLE SOLUTION

Theorem:

If a linear programming problem

Max. $Z = c\,x$ subject to $A\,x = b$, $x \geq 0$,

where $A = (\alpha_1, \alpha_2, \ldots, \alpha_N)$ is the coefficient matrix of order $m \times N$, $(N = m + n)$, has at least one feasible solution then it has at least one basic feasible solution also.

Proof:

Consider an arbitrary feasible solution of the given linear programming problem

$$x^* = (x_1, x_2, \ldots, x_N),\ x_i \geq 0.$$

Let us assume that k, $(k \leq N)$ variables in x* have positive values and the remaining $N - k$ variables are zero. We can also assume that the variables have numbered in such a way that the first k variables are non-zero.

Thus $x^* = (x_1, x_2, \ldots, x_k, 0, 0, \ldots, 0)$.

Also, we have $\sum_{i=1}^{k} x_i \alpha_i = b.$ ...(2)

Now two possibilities may arise:

The vectors $\alpha_1, \alpha_2, \ldots, \alpha_k$ may be either linearly independent or dependent.

Case 1.

If $\alpha_1, \alpha_2, \ldots, \alpha_k$ are linearly independent.

Any feasible solution for which the vectors a_i, i = 1, ... ,k associated with non-zero variables x_i, i = 1, ... ,k are L.I. is called a basic feasible solution. Thus x* is a B.F.S. which is also optimal.

Hence solution x* is degenerate if k < m and non-degenerate if k = m.

Case 2.

If $\alpha_1, \alpha_2, \ldots, \alpha k$ are linearly dependent and k > m. In this case shall reduce the number of non-zero variables step by step until we obtain a B.F.S. from feasible solution.

Since $\alpha_1, \alpha_2, \ldots, \alpha_k$ are linearly dependent therefore there exist scalars $\lambda_1, \lambda_2, \ldots \lambda_k$ such that

$$\lambda_1\ 1 + \lambda_2 + \alpha_2 + \ldots + \lambda_k\ \alpha_k = 0$$

$$\Rightarrow \sum_{o=1}^{k} \lambda_i \alpha_i = 0 \quad \text{...(3)}$$

and at least one $\lambda_1 \neq 0$.

We can assume that at least one λ_i is positive because if none is positive then we can multiply the equation (3) by – 1 and get positive λ_i.

Now, suppose

$$v = \max_{1 \le i \le k} \left(\frac{\lambda_i}{x_i} \right).$$

Obviously v is positive since $x_i \geq 0$, i = 1, 2, ..., k and at least one λ_i is positive. Multiplying both sides of (3) by 1/v and subtracting from (2), we get

$$\sum_{i=1}^{k}\left(x_i - \frac{\lambda_i}{v}\right)\alpha_i = b. \qquad \ldots(5)$$

(5) gives $x' = \left[x_1 - \frac{\lambda_1}{v}, x_2 - \frac{\lambda_2}{v}, \ldots, x_k - \frac{\lambda_k}{v}, 0, 0, \ldots, 0\right]$

a new solution of the matrix equation A x = b.

Also from (4) for i = 1, 2, ... ,k, we have

$$v \geq \frac{\lambda_i}{x_i} \text{ or } x_i \geq \frac{\lambda_i}{v} \quad \text{ or } \quad x_i - \frac{\lambda_i}{v} \geq 0.$$

Since all the components of x' are non-negative therefore x' is a feasible solution to the given L.P.P.

Also $v = \frac{\lambda_i}{x_i}$, for at least one i

i.e., $v - \frac{\lambda_i}{x_i} = 0$ therefore the new feasible solution x' cannot have more than k – 1 non-zero variables.

In this way we have derived a new feasible solution from the given optimal feasible solution which contains less number of non-zero variables. This solution is B.F.S. if the column vectors associated to non-zero variables in this new solution are L.I. If these associated vectors are not L.I., we shelf repeat the whole reduction procedure as explained above. Continuing in this way for finite number of times we will derive a solution in which columns corresponding to positive variables are L.I. *i.e.*, we will obtain a B.F.S. of the system.

Hence the theorem is proved.

Notes:

1. By applying the above maintained procedure we can obtain a basic feasible solution from any feasible solution.
2. *If the given feasible solution is optimal then basic feasible solution is also optimal.*

TO DETERMINE IMPROVED B.F.S. FROM A GIVEN B.F.S.

Theorem:

Let $x_B = B^{-1} b$ be a B.F.S. of a linear programming problem with $Z = c_B x_B$ as the value of the objective function. If for any column aj *in A, but not in B, the condition $c_j - Z_j > 0$ or $Z_j - c_j < 0$ holds and if at least one $y_{ij} > 0$, $i = 1, 2, \dots, m$, then it is possible to obtain a new B.F.S. by replacing one of the columns in B by α_j and if Z' is the new value of objective function then $Z' \geq Z$.*

If the initial B.F.S. $x_B = B^{-1} b$ is non-degenerate then $Z' > Z$.

Proof:

Consider the L.P.P.

Max. $Z = c\,x$, s.t. $A\,x = b$, $x \geq 0$,

where $A = (\alpha_1, \alpha_2, \dots, \alpha_N)$, $N = m + n$,

Basis matrix $B = (\beta_1, \beta_2, \dots, \beta_m)$.

Let $x_B\ (x_{B1}, x_{B2}, \dots, x_{Bm})$ be a B.F.S. of the given L.P.P.

Since the vectors $\beta_1, \beta_2, \dots, \beta_m$ are in the basis of A, therefore we can express α_j as a linear combination of b's.

$$\therefore \alpha_j = \sum_{i=1}^{m} y_{ij}\beta_i = y_{1j}\,\beta_1 + y_{2j}\,\beta_2 + \dots + y_{mj}\,\beta_m. \qquad \dots(1)$$

If $y_{rj} \neq 0$ then α_j can replace br in B and B is still a basis matrix.

Let $y_{rj} \neq 0$ then from (1), we have

$$\beta_r = \frac{1}{y_{rj}}\alpha_j - \frac{y_{1j}}{y_{rj}}\beta_1 - \dots - \frac{y_{(r+1)j}}{y_{rj}}\beta_{r+1} - \dots - \frac{y_{mj}}{y_{rj}}\beta_m$$

$$\Rightarrow \beta_r = \frac{1}{y_{rj}}\alpha_j - \sum_{i=1}^{m}\frac{y_{ij}}{y_{rj}}\beta_i\,. \qquad \dots(2)$$

Now, we have

$$Bx_B = b$$

$$\Rightarrow (\beta_1, \beta_2, \dots, \beta_r, \dots, \beta_m)\,(x_{B1}, x_{B2}, \dots, x_{Br}, \dots, x_{Bm}) = b$$

$$\Rightarrow \beta_1\,x_{B1} + \beta_2\,x_{B2} + \dots + \beta_r\,x_{Br} + \dots + \beta_m\,x_{Bm} = b$$

$$\Rightarrow \sum_{i=1}^{m}\beta_i x_{Bi} + \beta_r x_{Br} = b \qquad \dots(3)$$

Putting for br from (2) in (3), we get

$$\sum_{i=1}^{m}\beta_i\left[x_{Bi} - \frac{y_{ij}}{y_{rj}}x_{Br}\right] + \frac{x_{Br}}{y_{rj}}\alpha_j = b \qquad \dots(4)$$

$$\Rightarrow \sum_{i=1}^{m} x_{Bi}'\beta_i + x_{Br}'\alpha j = b, \qquad ...(5)$$

where $x_{Bi}' = x- - x_{Br}\dfrac{y_{ij}}{y_{rj}}$, $i = 1, 2, ..., m, i \neq r$

$$x_{Br}' = \frac{x_{Br}}{y_{rj}}, \; i = r$$

Comparing (3) and (5), we observe that new basic solution of A x = b is given by

$x'_B = (x'_{Bi}, x'_{Br})$, $i = 1, 2, ..., m, i \neq r$

where the values of x'_{Bi}, x'_{Br} are given by (6).

This basic solution will be feasible if we have

$$x_{Bi} - \frac{y_{ij}}{y_{rj}} x_{Br} \geq 0, \; i = 1, 2, ... m, \; i \neq r$$

and $\dfrac{x_{Br}}{y_{rj}} \geq 0, \; i = r.$ ...(7)

Since x_B is the initial B.F.S. we have

$x_{Br} \geq 0$, $r = 1, 2, ..., m$.

Thus we observe that (7) will hold only if

$y_{rj} > 0$ and $y_{ij} \leq 0$, $i = 1, 2, ..., m, i \neq r$.

If $y_{rj} > 0$ and $y_{ij} > 0$, then (7) is satisfied only if

$$\frac{x_{Bi}}{y_{ij}} - \frac{x_{Br}}{y_{rj}} \geq 0 \Rightarrow \frac{x_{Br}}{y_{rj}} \leq \frac{x_{Bi}}{y_{ij}}$$

$$\Rightarrow \frac{x_{Br}}{y_{rj}} = \min_{i}\left\{\frac{x_{Bi}}{y_{ij}}, y_{ij} > 0\right\} = v. \qquad ...(8)$$

Thus a new B.F.S. can be obtained from the initial B.F.S. by removing the column vector br of the basis matrix B by a_j if r is to be selected such that

$$v = \frac{x_{Br}}{y_{rj}} = \min_{i}\left\{\frac{x_{Bi}}{y_{ij}}, y_{ij} > 0\right\}.$$

If we have $v = 0$, which is possible only when $x_{Br} = 0$ then it means that the initial B.F.S. is degenerate.

Now we shall prove $Z' \geq Z$.

The value of the objective function for the initial B.F.S. x_B is

$$Z = c_B x_B$$

$$= (c_{B1}, c_{B2}, \ldots, c_{Bm})(x_{B1}, x_{B2}, \ldots, x_{Bm}) = \sum_{i=1}^{m} c_{Bi} x_{Bi}$$

Corresponding to the new B.F.S. x'_B, the value of the objective function is Z', so we have

$$Z' = \sum_{i=1}^{m} c'_{Bi}\, x'_{Bi}, \qquad \ldots(9)$$

where c_{Bi}' are the coefficients of the basic variables x_{Bi}' (i = 1, 2, ... ,m) in the objective function.

Obviously, $c_{Bi}' = c_{Bi}$, $i = 1, 2, \ldots, m,\ i \neq r$

and $\quad c_{Br}' = c_j$.

$\therefore$ we can write

$$Z' = \sum_{i=1}^{m} c_{Bi} x'_{Bi} + c_j x'_{Br}\,. \qquad \ldots(10)$$

Substituting for x_{Bi}', x_{Br}' from (6) in (10), we get

$$Z' = \sum_{i=1}^{m} c_{Bi}\left(x_{Bi} - x_{Br}\frac{y_{ij}}{y_{rj}}\right) + c_j \frac{x_{Br}}{y_{rj}}$$

$$= \sum_{i=1}^{m} c_{Bi}\left(x_{Bi} - x_{Br}\frac{y_{ij}}{y_{rj}}\right) + c_j \frac{x_{Br}}{y_{rj}}$$

since the term $c_{Bi}\left(x_{Bi} - x_{Br}\dfrac{y_{ij}}{y_{rj}}\right) = 0$ when i = r therefore it can be included in the summation without changing the value of Z'.

$$\therefore Z' = \sum_{i=1}^{m} c_{Bi} x_{Bi} - \frac{x_{Br}}{y_{rj}} \sum_{i=1}^{m} c_{Bi} y_{ij} + \frac{x_{Br}}{y_{rj}} c_j$$

$$= Z + \frac{x_{Br}}{y_{rj}}(c_j - Z_j), \text{ where } Z_j = \sum_{i=1}^{m} c_{Bi} y_{ij}$$

$$= Z + v\,(c_j - Z_j), \text{ from (8).} \qquad \ldots(11)$$

From (11), we observe that the new value of the objective function is equal to the original value of the objective function plus the quantity $v\,(c_j - Z_j)$.

We have $Z' \geq Z$ if $v\ (c_j - Z_j) \geq 0$.

Since $v \geq 0$ therefore $Z' \geq Z$ only if $c_j - Z_j \geq 0$.

Hence by choosing the vector α_j for which $c_j - Z_j > 0$ and at least one $y_{ij} > 0$, we obtain a new improved value of the objective function.

If the initial B.F.S. is non-degenerate then $v > 0$ and in that case $Z' > Z$.

Notes:

1. $B' = (\beta_1', \beta_2', \ldots, \beta_m')$ is he new non-singular matrix obtained from by replacing br with α_j and we have $\alpha_i' = \beta_i$, $i \neq r$ and $\beta_r' = \alpha_j$.
2. If $v = 0$ then the initial B.F.S. is degenerate. From (6) we have

$$x_{Bi}' = x_{Bi},\ i = 1, 2, \ldots, m,\ i \neq r$$

$$x_{Br}' = x_{Br} = 0,\ i = r.$$

Since the values of the variables common to the initial and new solutions do not change therefore the new B.F.S. is also degenerate.

CONDITIONS FOR THE EXISTENCE OF UNBOUNDED SOLUTIONS

In 5 we have proved that if for any column aj in A but not in B the condition $c_j - Z_j > 0$ holds and if at least one $y_{ij} > 0$, $i = 1, 2, \ldots, m$ then it is possible to find an improved B.F.S. Now the question arises : what will be the implication if for at least one a_j all $y_{ij} \leq 0$, $i = 1, 2, \ldots, m$. To get the answer of this question we shall prove the following theorem.

Theorem:

(Unbounded Solutions). If for any basic feasible solution $x_B = B^{-1} b$ to $A x = b$ there is some column α_j in A but not in B for which $c_j - Z_j > 0$ and $y_{ij} \leq 0$, $i = 1, 2, \ldots, m$ then if the objective function is to be maximized, the problem has an unbounded solution.

Proof:

Consider the linear programming problem

Max. $Z = c\,x$ s.t. $A\,x = b$, $x \geq 0$,

where $A = (\alpha_1, \alpha_2, \ldots \alpha_N)$. $N = m + n$,

basis matrix $B = (\beta_1, \beta_2, \ldots, \beta_m)$.

Let $x_B = (x_{B1}, x_{B2}, \ldots, x_{Bm})$ be a B.F.S. of the given L.P.P. Then we have

$$Bx_B = b \text{ or } \sum_{i=1}^{m} x_{Bi}\beta_i = b \quad \ldots(1)$$

and the corresponding value of the objective function is

$$Z = c_B x_B = \sum_{i=1}^{m} c_{Bi} x_{Bi} \quad \ldots(2)$$

Adding and subtracting $\lambda\alpha_j$ in (1), we get

$$\sum_{i=1}^{m} x_{Bi}\beta_i - \lambda\alpha_j + \lambda\alpha_j = b, \quad \ldots(3)$$

where l is some scalar.

Let $\alpha_j \in A$ s.t. $\alpha_j \notin B$.

Since the vectors $\beta_1, \beta_2, \ldots, \beta_m$ are in the basis of A therefore we can express α_j as the linear combination of b's

$$\therefore \alpha j = \sum_{i=1}^{m} y_{ij}\beta_i \quad \text{or} \quad -\lambda\alpha_j = -1 \sum_{i=1}^{m} y_{ij}\beta_i .$$

Substituting the above value of $-\lambda\alpha_j$ in (3), we get

$$\sum_{i=1}^{m} x_{Bi}\beta_i - \lambda \sum_{i=1}^{m} y_{ij}\beta_i + \lambda\alpha_j = b$$

$$\text{or } \sum_{i=1}^{m} (x_{Bi} - \lambda y_{ij})\beta_i + \lambda\alpha_j = b$$

Thus we obtain a new solution

$$x'_B = (x_{B1}', x_{B2}', \ldots, x_{Bm}', \lambda) \quad \ldots(4)$$

where $x_{Bi}' = x_{Bi} - \lambda y_{ij}$, $i = 1, 2, \ldots, m$.

When $\lambda > 0$, $x_{Bi} - \lambda y_{ij} \geq 0$ since $y_{ij} \leq 0$, $i = 1, 2, \ldots, m$ therefore (4) gives the feasible solution in which the number of positive variables is less than or equal to (m + 1). It may be less than (m + 1) because some$x_{Bi} - \lambda y_{ij}$ may be zero for some i. In case the number of positive variables in this solution is equal to (m + 1) then this solution will be non-basic feasible solution. If Z' is the new value of the objective function corresponding to this new solution, then we have

$$Z' = \sum_{i=1}^{m} c_{Bi}(x_{Bi} - \lambda y_{ij}) + c_j\lambda$$

or $$Z' = \sum_{i=1}^{m} c_{Bi} x_{Bi} + \lambda(c_j - c_{Bi} y_{ij}) = Z + \lambda\,(c_j - Z_j).$$

Since $c_j - Z_j > 0$ (given) therefore Z' can be made as large as we please by giving sufficiently large values to l. We know that a L.P.P. has an unbounded solution if the value of its objective function can be increased or decreased arbitrarily. Hence the given L.P.P. has an unbounded solution.

Note: If for same α_j, $Z_j - c_j > 0$ and $y_{ij} \le 0$, i = 1, 2, ... m then the L.P.P. has an unbounded solution if the objective function is to be minimized.

SOLVED EXAMPLES

Example 1:

If $x_1 = 1,\ x_2 = 0,\ x_3 = 1$ *be a feasible solution of the L.P.P.*

Min. $Z = 2x_1 + 3x_2 + 4x_3$

subject to $x_1 + x_2 + x_3 = 2$

$x_1 - x_2 + x_3 = 0,\ x_1, x_2, x_3 \ge 0$

then show that the given feasible solution is not basic.

Solution:

The given system of constraint equations can be written in matrix form A x = b

or $$\begin{bmatrix} 1 & 1 & 1 \\ 1 & -1 & 1 \end{bmatrix} \begin{bmatrix} x_1 \\ x_2 \\ x_3 \end{bmatrix} = \begin{bmatrix} 2 \\ 0 \end{bmatrix}.$$

$$\therefore \alpha_1 = \begin{bmatrix} 1 \\ 1 \end{bmatrix},\ \alpha_2 = \begin{bmatrix} 1 \\ -1 \end{bmatrix},\ \alpha_3 = \begin{bmatrix} 1 \\ 1 \end{bmatrix},\ b = \begin{bmatrix} 2 \\ 0 \end{bmatrix}.$$

The given feasible solution $x_1 = 1$, $x_2 = 0$, $x_3 = 1$ will not be basic if the vectors associated to non-zero variables are not linearly independent *i.e.*, if α_1 and α_3 are linearly dependent.

If we choose $\lambda_1 = -1$ and $\lambda_2 = 1$, we have

$$\lambda_1\,\alpha_1 + \lambda_2\,\alpha_2 = -1 \begin{bmatrix} 1 \\ 1 \end{bmatrix} + 1 \begin{bmatrix} 1 \\ 1 \end{bmatrix} = \begin{bmatrix} -1+1 \\ -1+1 \end{bmatrix} = \begin{bmatrix} 0 \\ 0 \end{bmatrix} = 0.$$

$\therefore$ the vectors α_1 and α_3 are linearly dependent.

Hence the given feasible solution is not basic.

Example 2:

If $x_1 = 2,\ x_2 = 3,\ x_3 = 1$ be a feasible solution of the linear programming problem

$$\text{Max. } Z = x_1 + 2x_2 + 4x_3$$

$$\text{s.t.} \quad 2x_1 + x_2 + 4x_3 = 11$$

$$3x_1 + x_2 + 5x_3 = 14$$

$$x_1, x_2, x_3 \geq 0.$$

then find a basic feasible solution from the given feasible solution.

Solution:

The given L.P.P. can be expressed as

$$\text{Max. } Z = x_1 + 2x_2 + 4x_3,$$

$$\text{s.t.} \quad A\,x = b$$

or $$\begin{bmatrix} 2 & 1 & 4 \\ 3 & 1 & 5 \end{bmatrix} \begin{bmatrix} x_1 \\ x_2 \\ x_3 \end{bmatrix} = \begin{bmatrix} 11 \\ 14 \end{bmatrix}$$

or $$\alpha_1\, x_1 + \alpha_2\, x_2 + \alpha 3\, x_3 = b, \qquad \text{...(1)}$$

where $\alpha_1 = \begin{bmatrix} 2 \\ 3 \end{bmatrix},\ \alpha_2 = \begin{bmatrix} 1 \\ 1 \end{bmatrix},\ \alpha_3 = \begin{bmatrix} 4 \\ 5 \end{bmatrix},\ b = \begin{bmatrix} 11 \\ 14 \end{bmatrix}.$

Since $x1 = 2$, $x_2 = 3$, $x_3 = 1$ is a feasible solution to the given L.P.P. therefore from (1)

$$2\alpha_1 + 3\alpha_2 + \alpha_3 = b.$$

Now the vectors $\alpha_1, \alpha_2, \alpha_3$ associated with non-zero variables x_1, x_2, x_3 will be linearly dependent if one of these vectors can be expressed as the linear combination of the remaining two vectors.

Let $$\lambda_1\, \alpha_1 + \lambda_2\, \alpha_2 = \alpha_3 \qquad \text{...(3)}$$

$$\Rightarrow \quad \lambda_1 \begin{bmatrix} 2 \\ 3 \end{bmatrix} + \lambda_2 \begin{bmatrix} 1 \\ 1 \end{bmatrix} = \begin{bmatrix} 4 \\ 5 \end{bmatrix}$$

$$\Rightarrow \quad \begin{bmatrix} 2\lambda_1 + \lambda_2 \\ 3\lambda_1 + \lambda_2 \end{bmatrix} = \begin{bmatrix} 4 \\ 5 \end{bmatrix}$$

$$\Rightarrow \quad 2\lambda_1 + \lambda_2 = 4,\ 3\lambda_1 + \lambda_2 = 5.$$

Solving these, we get

$$\lambda_1 = 1, \lambda_2 = 2.$$

Substituting the values of λ_1 and λ_2 in (3), we get

$$\alpha_1 + 2\alpha_2 = \alpha_3$$

$$\Rightarrow \alpha_1 + 2\alpha_2 - \alpha_3 = 0 \text{ or } \sum_{i=1}^{3} \lambda_i \alpha_i = 0$$

which gives $\lambda_1 = 1, \lambda_2 = 2, \lambda_3 = -1.$

Now we shall determine which of the three variables x_1, x_2, x_3 should be zero.

Using $v = \max_{1 \le i \le 3} \left(\dfrac{\lambda_i}{x_i} \right)$, we have

$$v = \max \left(\frac{\lambda_i}{x_1}, \frac{\lambda_2}{x_2}, \frac{\lambda_3}{x_3} \right),$$

$$\Rightarrow v = \max \left(\frac{1}{2}, \frac{2}{3}, -\frac{1}{1} \right) = \frac{2}{3} = \frac{\lambda_2}{x_2}.$$

Since $x_i - \dfrac{\lambda_i}{v} \ge 0$, therefore

$$\left(x_i - \frac{\lambda_1}{v}, x_2 - \frac{\lambda_2}{v}, x_3 - \frac{\lambda_3}{v} \right)$$

is the reduced feasible solution. We have

$$x_1 - \frac{\lambda_1}{v} = 2 - \frac{1}{2} = \frac{1}{2},$$

$$x_2 - \frac{\lambda_2}{v} = 0,$$

$$x_3 - \frac{\lambda_3}{v} = \frac{5}{2}.$$

$\therefore$ The new feasible solution is $x_1 = \dfrac{1}{2}$,

$$x_3 = 0, x_3 = \frac{5}{2}.$$

Now $\alpha_1 = \begin{bmatrix} 2 \\ 3 \end{bmatrix}$, and $\alpha_3 = \begin{bmatrix} 4 \\ 5 \end{bmatrix}$ are the vectors associated to non-zero variables x_1 and x_3.

Since α_1 and α_3 are linearly independent vectors therefore the new feasible solution

$$x_1 = \frac{1}{2}, x_2 = 0, x_3 = \frac{5}{2}$$

is the basic feasible solution.

Note: In order to find another basic feasible solution we proceed as follows:

Since $\alpha_1, \alpha_2, \alpha_3$ are L.D. therefore one of them can be expressed as the linear combination of the other two.

Let $\lambda_2 \alpha_2 + \lambda_3 \alpha_3 = \alpha_1$. ...(4)

Now proceed as in Example 2.

We have $\lambda_1 = -1, \lambda_2 = -2,$

$$\lambda_3 = 1, v = \frac{1}{1} = \frac{\lambda_3}{x_3}.$$

B.F.S. is $x_1 = 3, x_2 = 5, x_3 = 0$.

Example 3:

Consider the set of equations

$$5x_1 - 4x_2 + 3x_3 + x_4 = 3$$
$$2x_1 + x_2 + 5x_3 - 3x_4 = 0$$
$$x_1 + 6x_2 - 4x_3 + 2x_4 = 15,$$
$$x_1, x_2, x_3, x_4 \geq 0.$$

If $x_1 = 1, x_2 = 2, x_3 = 1, x_4 = 3$ is a feasible solution then find a basic feasible solution.

Solution:

The given set of equations can be expressed in the matrix form

$$\Rightarrow \begin{bmatrix} 5 & -4 & 3 & 1 \\ 2 & 1 & 5 & -3 \\ 1 & 6 & -4 & 2 \end{bmatrix} \begin{bmatrix} x_1 \\ x_2 \\ x_3 \\ x_4 \end{bmatrix} = \begin{bmatrix} 3 \\ 0 \\ 15 \end{bmatrix}$$

$$\Rightarrow \alpha_1 x_1 + \alpha_2 x_2 + \alpha_3 x_3 + \alpha_4 x_4 = b \qquad ...(1)$$

where $\alpha_1 = \begin{bmatrix} 5 \\ 2 \\ 1 \end{bmatrix}, \alpha_2 = \begin{bmatrix} -4 \\ 1 \\ 6 \end{bmatrix}, \alpha_3 = \begin{bmatrix} -3 \\ 5 \\ -4 \end{bmatrix}, \alpha_4 = \begin{bmatrix} 1 \\ -3 \\ 2 \end{bmatrix}, b = \begin{bmatrix} 3 \\ 0 \\ 15 \end{bmatrix}$.

Since $x_1 = 1, x_2 = 2, x_3 = 1, x_4 = 3$ is a feasible solution to the given set of equations, therefore from (1)

$\alpha_1 + 2\alpha_2 + \alpha_3 + 3\alpha_4 = b.$ ...(2)

Now the vectors α_1, α_2, α_3, α_4 associated with non-zero variables x_1, x_2, x_3, x_4 will be linearly dependent if one of these vectors can be expressed as the linear combination of the remaining three vectors.

Let $\lambda_2\,\alpha_2 + \lambda_3\,\alpha_3 + \lambda_4\,\alpha_4 = \alpha_1$ ...(3)

$$\Rightarrow \lambda_2 \begin{bmatrix} -4 \\ 1 \\ 6 \end{bmatrix} + \lambda_3 \begin{bmatrix} 3 \\ 5 \\ -4 \end{bmatrix} + \lambda_4 \begin{bmatrix} 1 \\ -3 \\ 2 \end{bmatrix} = \begin{bmatrix} 5 \\ 2 \\ 1 \end{bmatrix}$$

$$\Rightarrow \begin{bmatrix} -4\lambda_2 + 3\lambda_3 + \lambda_4 \\ \lambda_2 + 5\lambda_3 - 3\lambda_4 \\ 6\lambda_2 - 4\lambda_3 + 2\lambda_4 \end{bmatrix} = \begin{bmatrix} 5 \\ 2 \\ 1 \end{bmatrix}$$

$\Rightarrow -4\lambda_2 + 3\lambda_3 + \lambda_4 = 5$

$\lambda_2 + 5\lambda_3 - 3\lambda_4 = 2$

$6\lambda_2 - 4\lambda_3 + 2\lambda_4 = 1.$

Solving these, we get

$$\lambda_2 = \frac{22}{43},\ \lambda_3 = \frac{139}{86},\ \lambda_4 = \frac{189}{86}.$$

Substituting these values in (3), we get

$$\frac{22}{43}\alpha_2 + \frac{139}{86}\alpha_3 + \frac{189}{86}\alpha_4 = \alpha_1$$

$\Rightarrow 86\alpha_1 - 44\alpha_2 - 139\alpha_3 - 189\alpha_4 = 0$ ...(4)

$\Rightarrow \sum_{i=1}^{4} \lambda_i \alpha_i = 0$, which gives

$\lambda_1 = 86,\ \lambda_2 = -44,\ \lambda_3 = -139,\ \lambda_4 = -189.$

Using $v = \max_{1 \le i \le 4} \left(\frac{\lambda_i}{x_i}\right)$, we have

$$v = \max\left(\frac{\lambda_1}{x_1}, \frac{\lambda_2}{x_2}, \frac{\lambda_3}{x_3}, \frac{\lambda_4}{x_4}\right)$$

$$= \max\left(\frac{86}{1}, -\frac{44}{2}, -\frac{139}{1}, \frac{-189}{3}\right) = \frac{86}{1} = \frac{\lambda_1}{x_1}.$$

$\therefore$ the variable x_1 should be zero

i.e., the vector a1 should be eliminated.

Eliminating α_1 between (2) and (4), we get

$$\frac{44\alpha_2 + 139\alpha_3 + 189\alpha_4}{86} + 2\alpha_2 + \alpha_3 + 3\alpha_4 = b$$

$$\Rightarrow 0.\alpha_1 + \frac{216}{86}\alpha_2 + \frac{225}{86}\alpha_3 + \frac{447}{86}\alpha_4 = b$$

$$\therefore x_1 = 0,\ x_2 = \frac{216}{86},\ x_3 = \frac{225}{86},\ x_4 = \frac{447}{86}$$

is the new feasible solution.

Since the vectors $\alpha_2 = \begin{bmatrix} -4 \\ 1 \\ 6 \end{bmatrix}, \alpha_3 = \begin{bmatrix} 3 \\ 5 \\ -4 \end{bmatrix}, \alpha_4 = \begin{bmatrix} 1 \\ -3 \\ 2 \end{bmatrix}$

associated to non-zero variables are L.I. therefore the new feasible solution is the basic feasible solution.

EXERCISES

1. If $x_1 = 2$, $x_2 = 4$ and $x_3 = 1$ be a F.S. to the system of equations

   $2x_1 - x_2 + 2x_3 = 2,$

   $x_2 + 4x_2 = 18,$

   then find two basic feasible solutions.

2. (2, 1, 3) is a feasible solution of the set of equations

   $4x_1 + 2x_2 - 3x_3 = 1,$

   $6x_1 + 4x_2 - 5x_3 = 1.$

   Reduce the feasible solution to a b.f.s. of the set.

3. $x_1 = 1$, $x_2 = 1$, $x_3 = 1$, $x_4 = 0$ is a feasible solution to the system of equations

   $x_1 + 2x_2 + 4x_3 + x_4 = 7,$

   $2x_1 - x_2 + 3x_3 - 2x_4 = 4.$

   Reduce the feasible solution to two different basic feasible solutions.

4. State and prove the fundamental theorem of linear programming.

5. Show that if a linear programming problem has a feasible solution, it also has a basic feasible solution.